Flood and Tsunami Loads

Time-Saving Methods Using the 2018 IBC and ASCE/SEI 7-16

David A. Fanella,
Ph.D., S.E., P.E., F.ACI, F.ASCE, F.SEI

New York Chicago San Francisco
Athens London Madrid
Mexico City Milan New Delhi
Singapore Sydney Toronto

Library of Congress Cataloging-in-Publication Data

Names: Fanella, David Anthony, author.
Title: Flood and tsunami loads / David A. Fanella.
Description: New York : McGraw Hill Education, [2021] | "Time-Saving
    Methods Using the 2018 IBC and ASCE/SEI 7-16"—Verso title page. |
    Includes bibliographical references and index. | Summary: "A concise,
    visual guide for engineers working in flood and tsunami prone areas,
    this book will present explanations and workflows for the 20% of the
    building code that engineers use 80% of the time"—Provided by
    publisher.
Identifiers: LCCN 2020019678 | ISBN 9781260461503 (paperback ; acid-free
    paper) | ISBN 1260461505 (paperback ; acid-free paper) | ISBN
    9781260461510
Subjects: LCSH: Building, Stormproof—Standards. | Flood damage prevention.
    | Gust loads. | Tsunami resistant design. | Tsunami damage—Prevention.
Classification: LCC TH893 .F36 2020 | DDC 624.1/76—dc23
LC record available at https://lccn.loc.gov/2020019678

McGraw Hill books are available at special quantity discounts to use as premiums and sales promotions, or for use in corporate training programs. To contact a representative please visit the Contact Us page at www.mhprofessional.com.

Flood and Tsunami Loads:
Time-Saving Methods Using the 2018 IBC and ASCE/SEI 7-16

1 2 3 4 5 6 7 8 9   CD   25 24 23 22 21 20

ISBN 978-1-260-46150-3
MHID 1-260-46150-5

The pages within this book were printed on acid-free paper.

Sponsoring Editor	Project Manager	Production Supervisor
Ania Levinson	Parag Mittal, Cenveo® Publisher Services	Pamela A. Pelton
Editorial Supervisor		**Composition**
Donna M. Martone	**Copy Editor**	Cenveo Publisher Services
	Cenveo Publisher Services	
Acquisitions Coordinator		**Art Director, Cover**
Elizabeth Houde	**Proofreader**	Jeff Weeks
	Cenveo Publisher Services	

Contents

About the Author

David. A. Fanella is Senior Director of Engineering at the Concrete Reinforcing Steel Institute where his main responsibility is creating educational material for structural engineers, including publications, design aids, and webinars. He has over 30 years of experience in a wide variety of low-, mid-, and high-rise buildings and other structures and has authored numerous books and technical papers through the years, including two editions of Reinforced Concrete Structures, Analysis and Design. David is a licensed Structural Engineer and Professional Engineer in Illinois, and is a Fellow of the American Concrete Institute, the American Society of Civil Engineers, and the Structural Engineers Institute. He is active in many professional organizations, including membership on ASCE/SEI 7 and ACI Committees. He is also past President and past Board Member of the Structural Engineers Association of Illinois.

Preface

This publication provides structural engineers, educators, students, and other design professionals a concise, visual guide to the determination of structural loads for floods and tsunamis. The intent is to present the provisions in the 2018 *International Building Code* and ASCE/SEI 7-16 *Minimum Design Loads and Associated Criteria for Buildings and Other Structures* in a manner that is easy to understand and apply. This is achieved by utilizing step-by-step methods including numerous figures, tables, flowcharts, and design aids.

Examples in both inch-pound and metric (S.I.) units illustrate the proper application of the code provisions and follow the step-by-step methods outlined throughout the publication. Section, figure, table, and equation numbers from the code and this publication are given in the right-hand margin of the examples for easy reference.

In short, flood and tsunami loads can be determined simpler and faster using the procedures in this publication.

For further online information on this topic, please go to https://www.mhprofessional.com/FloodTsunamiLoads

David A. Fanella

CHAPTER 1

Introduction

1.1 Overview

The purpose of this publication is to assist in the proper determination of flood and tsunami loads in accordance with the 2018 edition of the *International Building Code©* (IBC©) [Reference 1; see Chapter 4 of this publication for a list of references] and the 2016 edition of ASCE/SEI 7 *Minimum Design Loads and Associated Criteria for Buildings and Other Structures* (Reference 2). The main goal is to streamline the load determination process by providing straightforward, step-by-step procedures enhanced by numerous design aids, figures, and flowcharts, which provide a roadmap through the numerous code requirements.

Design professionals will appreciate the simplicity and thoroughness of the content and will find the "how to" methods of load determination useful in everyday practice. Worked-out examples illustrate the proper application of the code requirements and follow the step-by-step procedures noted above; these examples are a valuable resource for individuals studying for licensing exams, undergraduate and graduate students, and others involved in structural engineering.

Readers interested in the background, history, and design philosophy of the code requirements for flood and tsunami loads can find detailed information and references in the commentary of Reference 2.

1.2 Scope

Throughout this publication, section numbers from the IBC are referenced as illustrated by the following: Section 1612 of the IBC is denoted as IBC 1612. Similarly, Section 5.4 of ASCE/SEI 7-16 is referenced as ASCE/SEI 5.4.

Chapter 2 contains methods to calculate design flood loads in accordance with IBC 1612 and ASCE/SEI Chapter 5. Included are procedures on how to determine hydrostatic, hydrodynamic, wave, and impact loads on structural components and foundations. Examples include the calculation of design flood loads for structures located in AE Zones (with riverine and coastal water sources), Coastal A Zones, and VE Zones.

Methods to calculate design tsunami loads in accordance with IBC 1615 and ASCE/SEI Chapter 6 are given in Chapter 3. Included are step-by-step procedures on how to determine tsunami inundation and runup at a site and how to calculate hydrostatic, hydrodynamic, and debris impact loads on the lateral force–resisting system and structural components of buildings. Examples illustrate the proper application of the code provisions for buildings located in tsunami design zones.

Chapter 4 contains the references cited in this publication.

Both inch-pound and S.I. units are used throughout this publication, including in the equations, figures, tables, flowcharts, and examples. In the examples, calculations are performed independently using both sets of units; in other words, the calculations are not performed in one set of units and then converted to the other. Thus, in some cases, the numerical results in inch-pound units do not "exactly" convert to the corresponding numerical results in S.I. units or vice versa.

CHAPTER 2
Flood Loads

2.1 Overview

This chapter contains methods to calculate design flood loads, F_a. All structures and portions of structures located in flood hazard areas (FHAs) must be designed and constructed to resist the effects of flood hazards and flood loads (IBC 1612.1).

The design and construction of buildings and structures located in FHAs, including Coastal High-Hazard Areas and Coastal A Zones, must be in accordance with Chapter 5 of ASCE/SEI 7 and ASCE/SEI 24 (Ref. 3).

In cases where a building or structure is located in more than one flood Zone or is partially located in a flood Zone, the entire building or structure must be designed and constructed according to the requirements of the more restrictive flood Zone.

Some structures will not be exposed to all the flood loads described in this chapter. The design professional is responsible for determining the applicable design flood loads applied to the structure.

2.2 Notation

A = surface area of a building or structure normal to the water flow, ft² (m²)

= projected area of debris accumulation into the flow, ft² (m²)

A_b = plan area of a building, ft² (m²)

A_c = area of a column, ft² (m²)

A_p = area of a pile, ft² (m²)

a = drag coefficient (shape factor)

BFE = base flood elevation, ft (m)

C_B = blockage coefficient for normal impact loads (see ASCE/SEI Table C5.4-3 or Figure C5.4-2)

C_D = coefficient of drag for breaking waves on vertical pilings and columns

= 1.75 for round piles or columns

= 2.25 for square piles or columns

C_D = depth coefficient for normal impact loads (see ASCE/SEI Table C5.4-2 or Figure C5.4-1)

C_I = importance coefficient for normal impact loads (see ASCE/SEI Table C5.4-1)

C_O = orientation coefficient for normal impact loads = 0.8

C_p = dynamic pressure coefficient for breaking wave loads on vertical walls (see ASCE/SEI Table 5.4-1)

D = pile or column diameter for circular sections, ft (m)

    = 1.4 times the width of a square pile or column, ft (m)

DFE = design flood elevation, ft (m)

d_b = basement depth, ft (m)

d_h = equivalent surcharge depth, ft (m)

d_s = stillwater flood depth, ft (m)

d_{sat} = depth of saturated soil, ft (m)

F = impact load, lb (kN)

    = drag force due to debris accumulation, lb (kN)

F_a = design flood load, lb (kN)

F_{buoy} = buoyancy load, lb (kN)

F_D = net wave load on vertical pilings and columns, lb (kN)

F_{dif} = load due to differential between water and soil pressures, lb (kN)

F_{dyn} = hydrodynamic load, lb (kN)

F_{nv} = horizontal component of a breaking wave load, lb/ft (kN/m)

F_{oi} = horizontal component of an obliquely incident breaking wave load, lb/ft (kN/m)

F_{sta} = hydrostatic load, lb (kN)

F_t = net breaking wave load per unit length acting on the vertical surface of a wall, lb/ft (kN/m)

f_{buoy} = buoyancy pressure, lb/ft^2 (kN/m^2)

G = existing ground elevation, ft (m)

GS = eroded ground elevation, ft (m)

g = acceleration due to gravity = 32.2 ft/s^2 (9.81 m/s^2)

H = load due to lateral earth pressure or groundwater pressure, lb (kN)

H_b = breaking wave height, ft (m)

NAVD = North American Vertical Datum of 1988

NGVD = National Geodetic Vertical Datum of 1929

n = number of piles or columns supporting a building or structure

P_{max} = maximum combined dynamic and static wave pressures, lb/ft^2 (kN/m^2)

R_{max} = maximum response ratio for impulsive loads (see ASCE/SEI Table C5.4-4)

T_n = natural period of a building, structure, or component, s

t = slab thickness, in. (mm)

V = design flood velocity, ft/s (m/s)

    = volume of floodwater displaced by a submerged object, ft^3 (m^3)

V_b = debris velocity, ft/s (m/s)

V_f = volume of a footing, ft³ (m³)

V_m = volume of a mat foundation, ft³ (m³)

V_w = volume of a wall, ft³ (m³)

W = debris impact weight, lb (kN)

w = width of a vertical component, ft (m)

     = width of a building or structure, ft (m)

α = vertical angle between a nonvertical wall surface and the horizontal

     = horizontal angle between the direction of wave approach and the vertical surface of a wall

γ_s = equivalent fluid weight of submerged soil and water, lb/ft³ (kN/m³)

γ_w = unit weight of water

     = 62.4 lb/ft³ (9.80 kN/m³) for fresh water

     = 64.0 lb/ft³ (10.05 kN/m³) for salt water

Δt = impact duration (time to reduce object velocity to zero), s

ρ = mass density of water = γ_w / g

     = 1.94 slugs/ft³ (1,000 kg/m³) for fresh water

     = 1.99 slugs/ft³ (1,026 kg/m³) for salt water

2.3   Terminology

Terminology used throughout this chapter related to flood loads is given in Table 2.1.

500-year flood elevation	The water surface elevation corresponding to the 0.2 percent annual chance flood (commonly referred to as the 500-year flood). This elevation is commonly specified in the Flood Insurance Study (FIS) or similar study.
A Zone	• A high-risk area within a Special Flood Hazard Area (SFHA) shown on flood insurance rate maps (FIRMs) not subjected to high-velocity wave action. • This Zone is also referred to as the Minimal Wave Action (MiWA) area where wave heights are less than 1.5 ft (0.46 m). • This Zone is designated on FIRMs as A, AE, A1 through A30, A99, AR, AO, and AH. See Figure C1-3 in ASCE/SEI 24.
Base flood	A flood having a 1 percent chance of being equaled or exceeded in any given year. This is commonly referred to as the extent of the 100-year floodplain.
Base flood elevation (BFE)	• Height to which flood waters will rise during passage or occurrence of the base flood relative to the datum. • The shape and nature of the floodplain (ground contours and the presence of any buildings, bridges, and culverts) are used to obtain the BFE along rivers and streams. • Along coastal areas, a BFE includes wave heights that are established considering historical storm and wind patterns. See Figure C1-3 in ASCE/SEI 24.

TABLE 2.1   Terminology Related to Flood Loads

Breakaway wall	Any type of wall subjected to flooding not required to provide structural support to a building or structure and designed and constructed to collapse under base flood or lesser conditions in such a way that it allows free passage of floodwaters and does not damage the structure or supporting foundation system. Loads on breakaway walls are given in ASCE/SEI 5.3.3.
Breaking wave height	Wave heights of depth-limited breaking waves (that is, wave heights limited by the depth of the water) in Coastal A Zones and V Zones.
Coastal A Zone	• An area within an SFHA landward of a V Zone or landward of an open coast without mapped V Zones (such as the shorelines of the Great Lakes). A schematic of a Coastal A Zone is given in Figure C1-1 of ASCE/SEI 24. Also see Figure C1-3 in ASCE/SEI 24. • This area is referred to as the Moderate Wave Action (MoWA) area by Federal Emergency Management Agency (FEMA) flood mappers and is between the Limit of Moderate Wave Action (LiMWA) and the V Zone. • The principal source of flooding is from astronomical tides, storm surges, seiches, or tsunamis and not from riverine flooding. • Wave forces and erosion potential should be taken into consideration when designing a structure for flood loads in such Zones. • Stillwater depth must be greater than or equal to 2.0 ft (0.61 m) and breaking wave heights are between 1.5 ft (0.46 m) and 3.0 ft (0.91 m) during the base flood. • This Zone is not delineated by FEMA as a separate Zone on FIRMs.
Coastal High Hazard Area (V Zone)	• An area within an SFHA with the following characteristics: 1. An area extending offshore to the inland limit of a primary frontal dune along an open coast. 2. An area subject to high-velocity wave action from storms or seismic sources. • This area (Zone) is designated on FIRMs as V, VE, or V1 through V30. See Figure C1-3 in ASCE/SEI 24.
Design flood	• The base flood identified on the community's FIRM where a community has chosen to adopt minimum National Flood Insurance Program (NFIP) building elevation requirements. • The flood corresponding to the area designated as a flood hazard area (FHA) on a community's flood hazard map or otherwise legally designated where a community has chosen to exceed minimum NFIP building elevation requirements.
Design flood elevation (DFE)	• The elevation of the design flood, including wave height (where applicable), relative to the datum specified on a community's flood hazard map. • For communities that have adopted the minimum NFIP requirements, the DFE is identical to the BFE. • The DFE is greater than the BFE in communities that have adopted requirements that exceed minimum NFIP requirements.
Digital flood insurance rate map (DFIRM)	A FIRM produced in a digital format.

TABLE 2.1   Terminology Related to Flood Loads (*Continued*)

Flood or flooding	A general and temporary condition of partial or complete inundation of normally dry land from the following (IBC 202): 1. The overflow of inland or tidal waters. 2. The unusual and rapid accumulation or runoff of surface waters from any source.
Flood exposure	Flood exposures are sources of flooding and are grouped into the following categories: 1. River, riverine, fluvial flooding: Rivers, lakes, manmade drainage channels, smaller watercourses overflowing due to upstream heavy rains, melting snow, and dam releases. 2. Alluvial fan flooding: Flooding that occurs in areas at the base of steep-sloped areas. As the water exists the steep area, it fans out into the flat areas in a random manner. 3. Coastal flooding: Oceans, bays, estuaries, and rivers affected by coastal waters overflowing due to abnormal high tides, coastal storms, high winds, or tsunamis. 4. Storm water flooding: Storm water flooding is caused by an accumulation of runoff on land and paved areas from rainfall before it enters a stream, river, body of water, or a manmade drainage system.
Flood hazard boundary map (FHBM)	The first flood risk map prepared by the FEMA for a community, which identifies FHAs based on the approximation of land areas in the community having a 1 percent or greater chance of flooding in any given year.
Flood hazard area (FHA)	An area defined as the greater of the following two areas (IBC 202): 1. The area within a floodplain subjected to a 1 percent or greater chance of flooding in any year (base flood); or, 2. The area designated as an FHA on a community's flood hazard map, or otherwise legally designated.
Flood insurance rate map	• The official map of a community prepared by the FEMA through the NFIP, which delineates both the special hazard areas and the risk premium Zones applicable to the community. • A FIRM shows FHAs along bodies of water where there is a risk of flooding by a Base Flood. Examples of coastal and riverine FIRMs are given in Figure C1-3 of ASCE/SEI 24.
Flood Insurance Study	• A report prepared by the FEMA to document the examination, evaluation, and determination of flood hazards. • Included in an FIS are the FIRM, the flood boundary and floodway map (FBFM), the BFE, and supporting technical data.
Floodway	A channel of a river, creek, or other watercourse and adjacent land areas that must be reserved in order to discharge the base flood waters without cumulatively increasing the water surface elevation by more than a designated height. A floodway schematic for an SFHA is given in Figure C1-2 of ASCE/SEI 24.
Freeboard	Additional depth of water above the BFE approved by a local jurisdiction.

TABLE 2.1 Terminology Related to Flood Loads (*Continued*)

High-risk flood hazard area	FHA where one or more of the following hazards are known to occur: • Alluvial fan flooding • Flash floods • Mudslides • Ice jams • High-velocity flows • High-velocity wave action • Breaking wave heights greater than or equal to 1.5 ft (0.46 m) • Erosion
Limit of Moderate Wave Action	• The inland limit of the area expected to receive 1.5-ft (0.46-m) or greater breaking waves during the base flood. • The boundary between the MoWA and MiWA areas.
Minimal Wave Action Area	• The area between the LiMWA and the landward limit of an A Zone. • Wave heights are less than 1.5 ft (0.46 m) during the base flood.
Moderate Wave Action Area	The area between the LiMWA and the V Zone (see the definition for Coastal A Zone).
National Flood Insurance Program (NFIP)	A program administered by the FEMA that enables property owners in participating communities to purchase federally backed flood insurance as financial protection against flood losses in exchange for state and community floodplain management regulations that reduce future flood damage.
Special flood hazard area (SFHA)	The land area subject to flood hazards, which are shown on a FIRM or other flood hazard map as A Zones (A, AE, A1 through A30, A99, AR, AO, AH) or V Zones (V, VO, VE, or V1 through V30). See Figure C1-3 in ASCE/SEI 24.
Stillwater elevation (SWEL)	• In riverine and lake areas, the BFE published in the FIS and shown on the FIRM, unless the local jurisdiction has adopted a more severe design flood. • In coastal areas, the average water level including waves, which is published in the FIS. • The SWEL must be referenced to the same datum used in establishing the BFE and the DFE.
Stillwater flood depth	The vertical distance between the eroded ground elevation and the SWEL.
V Zone	See the definition for Coastal High-Hazard Area.
X Zone	• Formerly identified as B and C Zones on older FIRMs, an area outside of the flood hazard Zone. • X Zones are identified as follows: 1. Shaded X Zone (formerly B Zones): Identifies areas subject to flooding having a 0.2 percent probability of being equaled or exceeded during any given year (500-year flood). 2. Unshaded X Zone (formerly C Zones): Identifies areas of minimal flood risk where the annual exceedance probability of flooding is less than 0.2 percent.

TABLE 2.1   Terminology Related to Flood Loads (*Continued*)

2.4   Procedure to Determine Flood Loads, F_a

A step-by-step procedure to determine flood loads, F_a, is given in Fig. 2.1. It is assumed the building or structure is located in an FHA in accordance with the flood hazard map of the local jurisdiction. The section, table, and figure numbers of this publication referenced in Fig. 2.1 contain information on the quantities needed to calculate F_a.

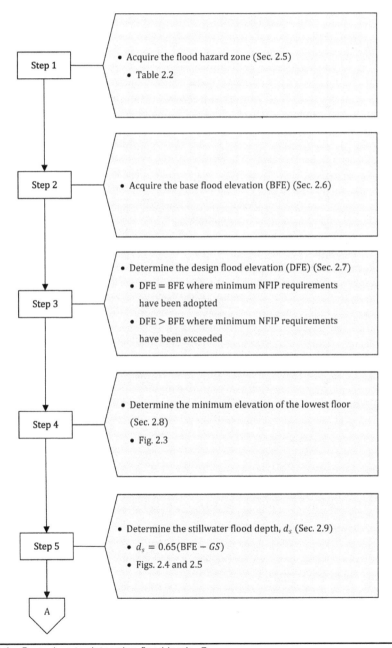

FIGURE 2.1   Procedure to determine flood loads, F_a.

14   Chapter Two

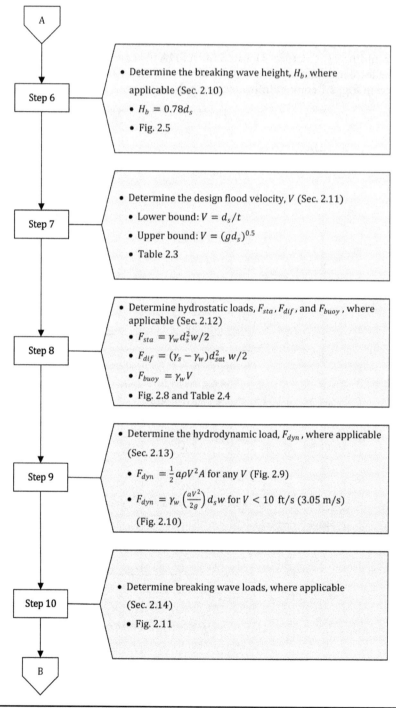

FIGURE 2.1   (Continued)

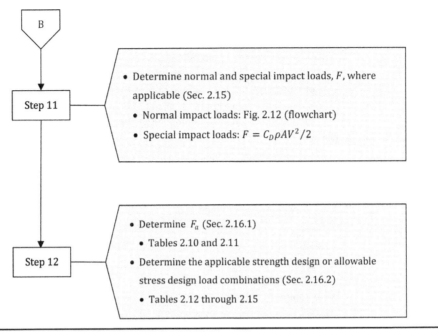

- Determine normal and special impact loads, F, where applicable (Sec. 2.15)
  - Normal impact loads: Fig. 2.12 (flowchart)
  - Special impact loads: $F = C_D \rho A V^2 / 2$

- Determine F_a (Sec. 2.16.1)
  - Tables 2.10 and 2.11
- Determine the applicable strength design or allowable stress design load combinations (Sec. 2.16.2)
  - Tables 2.12 through 2.15

FIGURE 2.1   *(Continued)*

2.5   Flood Hazard Zones

The flood hazard Zone must be determined for any structure located in an FHA. The FEMA determines flood hazards at a given site based on the following:

1. Anticipated flood conditions, including the stillwater elevation, wave setup, wave runup and overtopping, and wave propagation during the base flood event (that is, the flood level having a 1 percent chance of being equaled or exceeded in any given year).

2. Potential for storm-induced erosion during the base flood event.

3. Physical characteristics of the floodplain, including existing development and vegetation.

4. Topographic and bathymetric (underwater topography) information.

Flood hazards and water surface elevations are calculated by FEMA computer models and the results from these analyses are used to map flood hazard Zones and base flood elevations (BFEs).

FEMA flood hazard Zones are given in Table 2.2 (also see Table 2.1). Coastal A Zones are not given by FEMA as a separate Zone on FIRMs; a Coastal A Zone designation is used to facilitate application of load combinations in Chapter 2 of ASCE/SEI 7 (see Sec. 2.16 of this publication).

The flood insurance rate map (FIRM) for a given area contains both the special hazard areas and the risk premium Zones applicable to the community. A generic FIRM panel for an area located along a coastline is given in Fig. 2.2. The elevations in parentheses below

Zone	Description
Moderate- to Low-Risk Areas—X Zones	
X	These Zones identify areas outside of the flood hazard area (FHA). • Shaded X Zone: areas subject to flooding having a 0.2 percent probability of being equaled or exceeded during any given year. • Unshaded X Zone: areas of minimal flood risk where the annual exceedance probability of flooding is less than 0.2 percent.
High-Risk Areas—A Zones	
A	• Detailed analyses are not performed for such areas. • No depths or BFEs are shown within these Zones.
AE	• BFEs are provided. • AE Zones are used on new format FIRMs instead of A1 through A30 Zones.
A1 through A30	• These are known as numbered A Zones (for example, A7 or A14). • Old-format FIRMs show a BFE for these Zones.
A99	• Areas protected by a Federal flood control system where construction has reached specified legal requirements. • No depths or BFEs are shown within these Zones.
AR	Areas with a temporarily increased flood risk due to the building or restoration of a flood control system (such as a levee or a dam).
AO	• River or stream FHAs and areas with a chance of shallow flooding, usually in the form of sheet flow, with an average depth ranging from 1 to 3 ft (0.30 to 0.91 m). • Average flood depths derived from detailed analyses are shown within these Zones.
AH	• Areas of shallow flooding, usually in the form of a pond, with an average depth ranging from 1 to 3 ft (0.30 to 0.91 m). • BFEs are derived from detailed analyses and are shown at selected intervals within these Zones.
High-Risk Areas—Coastal (V Zones)	
V	• Coastal areas with a 1 percent or greater chance of flooding and an additional hazard associated with storm waves. • No BFEs are shown within these Zones.
VE, V1 through 30	• Coastal areas with a 1 percent or greater chance of flooding and an additional hazard associated with storm waves. • BFEs are derived from detailed analyses and are shown at selected intervals within these Zones.

TABLE 2.2   FEMA Flood Hazard Zones

the Zone designation are the BFEs (in feet) with respect to either the North American Vertical Datum of 1988 (NAVD 88 or NAVD) or the National Geodetic Vertical Datum of 1929 (NGVD 29 or NGVD); see Sec. 2.6 of this publication. The FIRM identifies which datum was used to establish the elevation. FIRM panels can be obtained from Ref. 4 and flood-related information needed to calculate flood loads can be acquired from Ref. 5.

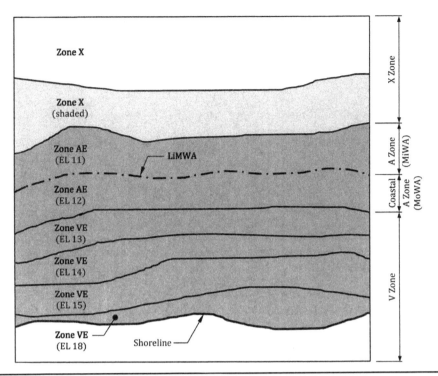

FIGURE 2.2 Sample FIRM for an area along a coastline.

2.6 Base Flood Elevation

The BFE is the height to which flood waters are anticipated to rise during a passage or occurrence of the base flood relative to the datum. In addition to showing the extent of the flood hazard Zones, FIRMs and DFIRMs also give the BFE, which is rounded to the nearest whole foot. The BFE is typically located in parentheses below the flood hazard Zone on the flood hazard map (see Fig. 2.2). For riverine areas, the BFE is usually located adjacent to river cross-section locations.

Modern FIRMs contain BFEs based on the North American Vertical Datum of 1988 (NAVD 88). Historically, the most common vertical datum used by FEMA has been the National Geodetic Vertical Datum of 1929 (NGVD 29), and many existing FIRMs provide elevation values based on this old datum. Where the datums on two or more sources are different (for example, the elevations on a FIRM are based on NAVD 88 and the elevations on an elevation certificate are based on NGVD 29), the elevations must be converted to the same datum before they are used subsequently. Differences between two datums vary from location to location and FEMA provides guidelines on how to calculate the conversions (see Ref. 6).

2.7 Design Flood Elevation

The design flood elevation (DFE) is used in the determination of flood loads and minimum elevation requirements of the lowest floor of a building or structure (see Sec. 2.8 of this publication).

The DFE is the elevation of the design flood including wave height (where applicable) and is equal to the following:

- For communities that have adopted minimum NFIP requirements:

$$DFE = BFE$$

- For communities that have adopted requirements exceeding those from the NFIP:

$$DFE > BFE$$

To account for uncertainties in the determination of flood elevations, local jurisdictions may require an additional depth of water above the BFE, which is defined as freeboard. Freeboard is essentially an additional factor of safety that provides an increased level of flood protection, which could reduce flood insurance premiums. Note that freeboard should not be included in the determination of the stillwater flood depth (see Sec. 2.9 of this publication) but is used instead to raise the building to a level higher than the BFE level (see Sec. 2.8 of this publication).

2.8   Minimum Elevation of Lowest Floor

IBC 1612.4 and ASCE/SEI 24 Sections 2.3 and 4.4 require the lowest floor in a building or structure be elevated to or above the minimum elevations identified in ASCE/SEI 24 Table 2-1 for A Zones and Table 4-1 for Coastal A Zones and V Zones; the minimum elevations are based on the flood design class, the BFE, the DFE, and the 500-year flood elevation (see Fig. 2.3). Flood design classes are defined in ASCE/SEI 24 Table 1-1 and are similar to the risk categories in IBC Table 1604.5 and in ASCE/SEI Table 1.5-1.

For buildings located in an A Zone, the minimum elevation is to the top of the lowest floor whereas for buildings in coastal A Zones or V Zones, the minimum elevation is to the bottom of the lowest supporting horizontal structural member of the lowest floor.

Enclosed areas used solely for parking of vehicles, building access, or storage are permitted to be below the DFE of elevated buildings in an A Zone provided the requirements in ASCE/SEI 24 Section 2.7 for such enclosures are satisfied (ASCE/SEI 24 Section 2.3). According to the exception in ASCE/SEI 24 Section 2.3, the lowest floor (including basements) in a nonresidential building or in nonresidential portions of a mixed-use building in an A Zone is permitted to be located below the minimum elevations in ASCE/SEI 24 Table 2-1 only if the dry floodproofing requirements in ASCE/SEI Section 6.2 are satisfied. There are no similar exceptions for buildings located in a Coastal A Zone or V Zone; all buildings must satisfy the elevation requirements in ASCE/SEI 24 Table 4-1 (ASCE/SEI 24 Section 4.4).

2.9   Stillwater Flood Elevation (SWEL) and Stillwater Flood Depth, d_s

Unless a local jurisdiction has adopted a more severe design flood, the SWEL in riverine and lake areas is equal to the BFE published in the FIS and shown on the FIRM (see Fig. 2.4 for the case where DFE = BFE). In coastal areas, the SWEL is the average water level including wave setup (see Fig. 2.5 for the case where DFE = BFE) and is published in the FIS (such values are not shown on FIRMs). The SWEL must be referenced to the same datum used in determining the BFE and the DFE.

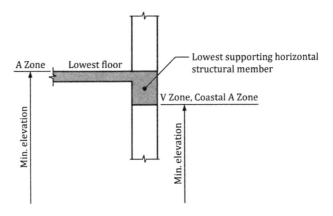

Flood Design Class	Minimum Elevation	
	A Zone	V Zone Coastal A Zone
1	DFE	DFE
2	Higher of { BFE + 1 ft / DFE	Higher of { BFE + 1 ft / DFE
3	Higher of { BFE + 1 ft / DFE	Higher of { BFE + 2 ft / DFE
4	Higher of { BFE + 2 ft / DFE / 500-yr FE	Higher of { BFE + 2 ft / DFE / 500-yr FE

1 ft = 0.3048 m
See ASCE/SEI 24 Table 1-1 for descriptions of Flood Design Class.
FE = flood elevation
See footnotes to ASCE/SEI 24 Table 2-1, which are applicable to A Zones.

FIGURE 2.3 Minimum elevation of the lowest floor in a building or structure.

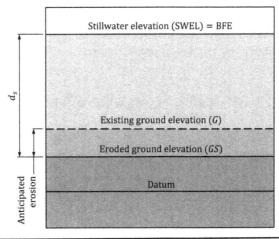

FIGURE 2.4 Stillwater flood elevation and stillwater flood depth for riverine and lake areas.

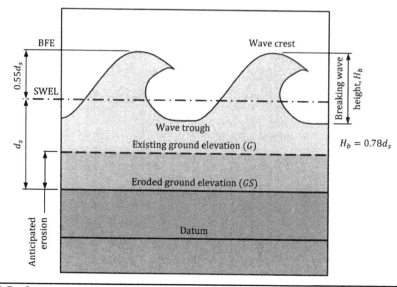

FIGURE **2.5**   Stillwater flood elevation and stillwater flood depth for coastal areas with waves.

The stillwater flood depth, d_s, is defined as the vertical distance between the SWEL and the eroded ground level. In coastal areas, d_s must include the effects of wave setup, which is the increase in the stillwater surface near the shoreline due to the presence of breaking waves (see Sec. 2.10 of this publication). Wave setup was typically not included in FIS reports and FIRMs prior to 1989. It is important to check the FIS report regarding wave setup: If wave setup is included in the BFE but not in the SWEL, wave setup must be added before calculating d_s, the design wave height, the design flood velocity, flood loads, and localized scour. If it has been determined that wave setup is included in the SWEL, d_s and the other parameters can be calculated based on that SWEL.

Floods may erode soil in both riverine and coastal areas. The existing ground elevation must be lowered if erosion has occurred in the vicinity of the site; erosion causes an increase in d_s, which in turn, increases the required design flood loads. Using historical information, contacting local geotechnical engineers, or contacting the local jurisdiction may provide some insight into the anticipated amount of erosion. Reference 7 provides guidance on estimating erosion in coastal areas.

2.10   Breaking Wave Height, H_b, and Wave Runup

Breaking wave height, H_b, is the vertical distance between the crest and the trough of a wave. In shallow waters, H_b depends on d_s and is typically determined by ASCE/SEI Equation (5.4-2):

$$H_b = 0.78d_s \tag{2.1}$$

Where steep slopes exist immediately seaward of a building, wave heights can exceed those determined by Eq. (2.1); in such cases, a reasonable alternative is to set $H_b = 1.0d_s$.

Because wave forms in shallow waters are distorted, the crest and trough of a wave are not equidistant from the SWEL. For NFIP mapping purposes, it is assumed that 70 percent of the wave height H_b is above the SWEL. Thus, the maximum elevation of a breaking wave crest above the SWEL is equal to $0.7 \times 0.78 d_s = 0.55 d_s$ (see Fig. 2.5). Based on this assumption, the following relationship can be established:

$$\text{BFE} - GS = d_s + 0.55 d_s = 1.55 d_s$$

or

$$d_s = 0.65\,(\text{BFE} - GS) \qquad\qquad (2.2)$$

ASCE/SEI Equation (5.4-3) is the same as Eq. (2.2) except G is used in ASCE/SEI Equation (5.4-3) instead of GS where G is equal to the lowest eroded ground elevation.

Equation (2.2) is applicable in coastal FHAs where the ground slopes up gradually from the shoreline, and there are few obstructions seaward of a building, such as vegetation or other buildings. The BFE shown on the FIRM in such areas is approximately equal to $GS + 1.55 d_s$. This case is illustrated in Fig. 2.6, which also includes the associated flood hazard Zones.

In areas with steeply sloping shorelines, water can rush up the surface, which could result in flood elevations higher than those based on breaking waves. This phenomenon is commonly referred to as wave runup. In coastal FHAs where this can occur, the BFE shown on the FIRM is equal to the highest elevation reached by the water (see Fig. 2.7).

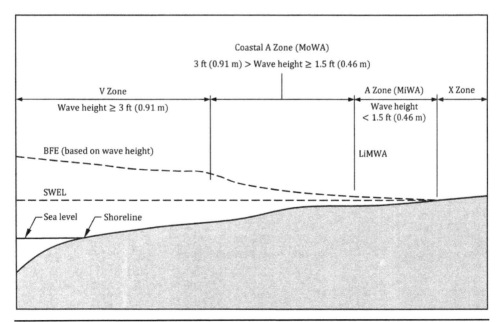

FIGURE 2.6 BFE and SWEL at locations with a gradually sloping shoreline.

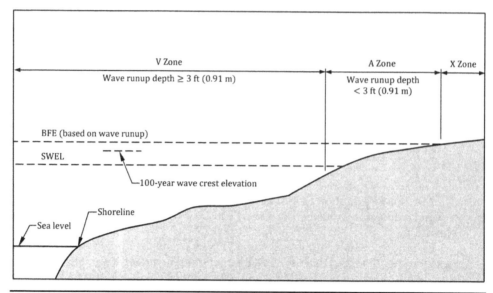

FIGURE 2.7 BFE and SWEL at locations with a steeply sloping shoreline.

2.11 Design Flood Velocity, *V*

Design flood velocity, V, is used in the determination of hydrodynamic flood loads (see Sec. 2.13 of this publication). Due to the inherent uncertainties of estimating design flood velocities, especially in coastal FHAs, the lower and upper bound velocities in Table 2.3 can be used, which are based on d_s. Guidelines on when to use the lower and upper bound velocities are also given in Table 2.3.

Lower bound	$V = d_s/t$	To be used where one or more of the following conditions exist: • Where the site is distant from the flood source • In an A Zone • On flat or gently sloping terrain • Where the building is not affected by other buildings or obstructions
Upper bound	$V = (gd_s)^{0.5}$	To be used where one or more of the following conditions exist: • If the site is near the flood source • In a V Zone • In an AO Zone adjacent to a V Zone • In an A Zone subject to velocity flow and wave action • On steeply sloping terrain • Where the building is adjacent to other large buildings or obstructions that will confine or redirect floodwaters and increase local flood velocities

TABLE 2.3 Design Flood Velocity, *V*

In the equations for design flood velocity, V is in ft/s (m/s), d_s is in ft (m), g = acceleration due to gravity = 32.2 ft/s² (9.81 m/s²), and $t = 1$ s.

It is important to note that the design velocities in Table 2.3 are not applicable to tsunami events. For estimating flood velocities during tsunamis, see Chap. 3 of this publication.

2.12   Hydrostatic Loads

2.12.1   Overview

Hydrostatic loads occur when stagnant or slow-moving water [velocity less than 5 ft/s (1.52 m/s); see ASCE/SEI C5.4.2] comes into contact with a foundation, building, or building component. The water can be above or below the ground surface. Such loads can be subdivided into lateral loads, vertical downward loads, and vertical upward loads (uplift or buoyancy). In general, a hydrostatic load is equal to the water pressure times the surface area on which the pressure acts.

2.12.2   Lateral Hydrostatic Loads

For vertical components of a building above ground, the horizontal hydrostatic load, F_{sta}, is the resultant force of the hydrostatic water pressure, which is equal to zero at the surface of the water and increases linearly to $\gamma_s d_s$ at the stillwater depth. The horizontal hydrostatic load can be determined by the following equation:

$$F_{sta} = \frac{\gamma_w d_s^2 w}{2} \tag{2.3}$$

In this equation, γ_w is the unit weight of water, which is equal to 62.4 lb/ft³ (9.80 kN/m³) for fresh water and 64.0 lb/ft³ (10.05 kN/m³) for saltwater, and w is the width of the vertical building component in ft (m). The load F_{sta} acts perpendicular to the surface and is located $2d_s/3$ below the BFE. The hydrostatic loads on the above-ground foundation walls of a nonresidential building in an A Zone are illustrated in the upper part of Fig. 2.8 where it is assumed that the dry floodproofing requirements of ASCE/SEI 24 Section 6.2 are satisfied, which permits the lowest floor to be below the minimum elevation in ASCE/SEI 24 Table 2-1 (see Sec. 2.8 of this publication). When calculating hydrostatic loads on surfaces exposed to free water, the design depth must be increased by 1.0 ft (0.30 m) [ASCE/SEI 5.4.2]. This is equivalent to increasing d_s by 1.0 ft (0.30 m) when calculating F_{sta} by Eq. (2.3).

For vertical components of a building above and below ground (such as basement walls), the total lateral hydrostatic pressure consists of the pressure due to the standing floodwater and the pressure due to the saturated soil (see ASCE/SEI 3.2.1 and the lower part of Fig. 2.8):

- Lateral hydrostatic load due to the water, which is based on the distance from the BFE to the top of the footing:

$$F_{sta} = \frac{\gamma_w (d_s + d_b + t)^2 w}{2}$$

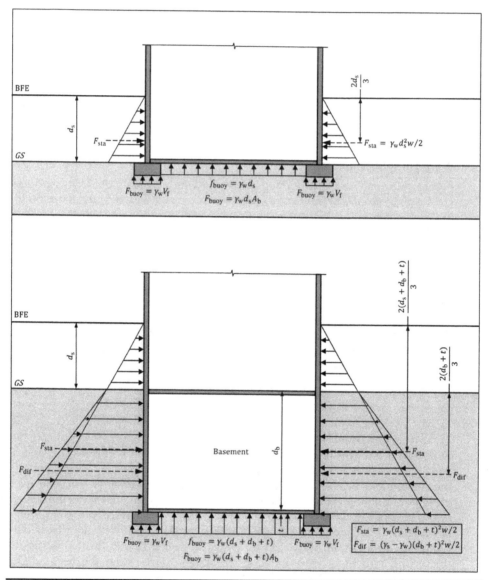

Figure 2.8 Hydrostatic loads on nonresidential buildings located in an A Zone where the dry floodproofing requirements of ASCE/SEI 24 Section 6.2 are satisfied.

- Lateral hydrostatic load due to the differential between the water and soil pressures, which is based on the depth of the saturated soil from the adjacent grade to the top of the footing:

$$F_{\text{dif}} = \frac{(\gamma_s - \gamma_w)(d_b + t)^2 w}{2}$$

Soil Type (Symbols)	γ_s, lb/ft³ (kN/m³)
Clean sand and gravel (GW, GP, SW, SP)	75 (11.78)
Dirty sand and gravel of restricted permeability (GM, GM-GP, SM, SM-SP)	77 (12.10)
Stiff residual silts and clays, silty fine sands, clayey sands and gravels (CL, ML, CH, MH, SM, SC, GC)	82 (12.88)
Very soft to soft clay, silty clay, organic silt and clay (CL, ML, OL, CH, MH, OH)	106 (16.65)
Medium to stiff clay deposited in chunks and protected from infiltration (CL, CH)	142 (22.31)

TABLE 2.4 Equivalent Fluid Weight of Submerged Soil and Water, γ_s

In these equations, d_b is the height of the basement and t is the thickness of the basement slab. In the equation for F_{dif}, γ_s is the equivalent fluid weight of submerged soil and water, which is based on soil type. Values of γ_s are given in Table 2.4 (see Ref. 8; it is always good practice to confirm γ_s for the site with the local jurisdiction or with other references that contain measured soil properties). The soil types and group symbols in Table 2.4 are from the United States Department of Agriculture (USDA) unified soil classification system and are defined in Table 2.5.

Soil Type	Group Symbol	Description
Gravels	GW	Well-graded gravels and gravel mixtures
	GP	Poorly graded gravel-sand-silt mixtures
	GM	Silty gravels, gravel-sand-silt mixtures
	GC	Clayey gravels, gravel-sand-clay mixtures
Sands	SW	Well-graded sands and gravelly sands
	SP	Poorly graded sands and gravelly sands
	SM	Silty sands, poorly graded sand-silt-mixtures
	SC	Clayey sands, poorly graded sand-clay mixtures
Fine grain silt and clays	ML	Inorganic silts and clayey silts
	CL	Inorganic clays of low to medium plasticity
	OL	Organic silts and organic silty clays of low plasticity
	MH	Inorganic silts, micaceous or fine sands or silts, elastic silts
	CH	Inorganic clays of high plasticity, fine clays
	OH	Organic clays of medium to high plasticity

TABLE 2.5 Soil Type Definitions Based on USDA Unified Soil Classification System

2.12.3   Vertical Hydrostatic Loads (Buoyancy Loads)

In addition to lateral hydrostatic loads, vertical structural elements above and below ground are subjected to upward hydrostatic loads, which are commonly referred to as buoyancy loads. In general, the buoyancy load, F_{buoy}, is equal to the following:

$$F_{\text{buoy}} = \gamma_w V \qquad (2.4)$$

where V is the volume of floodwater displaced by the submerged object in ft³ (m³). For example, the buoyancy load on a column above ground and supporting a building or structure is equal to $F_{\text{buoy}} = \gamma_w d_s A_c$ where A_c is the area of the column. This load acts in the upward direction and is perpendicular to A_c.

Illustrated in Fig. 2.8 are the hydrostatic uplift loads (buoyancy loads) on the lowest floor slab in each building assuming the soil below the slab becomes saturated by the floodwaters (see ASCE/SEI 3.2.2). The buoyancy pressure, f_{buoy}, on the underside of the slab in the upper part of Fig. 2.8 is equal to $\gamma_w d_s$. The total buoyancy load on the slab is $F_{\text{buoy}} = \gamma_w d_s A_b$ where A_b is the plan area of the building. It is evident that $d_s A_b$ is the volume of floodwater displaced by the submerged portion of the building. The buoyancy pressure on the underside of the slab in the lower part of Fig. 2.8 is equal to $f_{\text{buoy}} = \gamma_w (d_s + d_b + t)$ and the total buoyancy load in this case is $F_{\text{buoy}} = \gamma_w (d_s + d_b + t) A_b$. In cases where the slab is isolated from the building structure (which is a typical slab-on-grade), the slab is subjected to the buoyancy pressure and overall buoyancy (uplift) on the building need not be considered. However, if the slab is connected to the building structure and is designed to resist the buoyancy pressure, uplift on the building must be considered.

Buoyancy loads due to the volume of water displaced by the portions of the foundation walls below the DFE (that is, the submerged portions of the basement/foundation walls) and the wall footings must also be considered (see Fig. 2.8 for the buoyancy loads on the footings). The buoyancy load on a footing is equal to $F_{\text{buoy}} = \gamma_w V_f$ where V_f is the volume of the footing. Similarly, the buoyancy load per unit length of a foundation or basement wall is equal to γ_w times the thickness of the wall times the depth of the submerged portion of the wall.

To prevent flotation of a building during a design flood event, the total buoyancy load, which can be relatively large for buildings that have dry floodproofed basements and structural slabs, is resisted by the weight of the building (including basement walls where applicable) and the weight of the foundations. Because floods can occur at any time, any transient loads, such as the live loads, should not be included to resist uplift loads. Where the weights of the building and foundations are not sufficient, other structural measures (such as tension piles or additional mass) must be provided. Alternatively, vent openings in the foundation walls can be incorporated to alleviate the buoyancy loads.

Cases where hydrostatic loads do not need to be considered are given in Table 2.6 based on flood hazard area.

2.13   Hydrodynamic Loads

Hydrodynamic loads are caused by floodwaters flowing past a fixed object, such as a building or a supporting element (for example, a pile or a column). Like wind flowing around a building, the loads produced by moving water include an impact load on the

Flood Hazard Area	Requirements	Notes
A Zones	Foundation walls and exterior enclosure walls below the DFE that do not meet the dry floodproofing requirements in ASCE/SEI 24 Section 6.2 must contain openings to allow for automatic entry and exit of floodwaters during design flood conditions (ASCE/SEI 24 Section 2.7.1).	For walls with flood openings, lateral hydrostatic loads are equal to zero because the external and internal lateral hydrostatic pressures are balanced.
V Zones and Coastal A Zones	• The bottom of the lowest horizontal structural member of the lowest floor must be elevated in accordance with the minimum requirements of ASCE/SEI 24 Table 4-1. • Buildings must be supported on open foundations (such as piles, posts, piers, or columns) free of obstructions and that will not restrict or eliminate free passage of high-velocity floodwaters and waves (ASCE/SEI 24 Section 4.5.1).	Lateral hydrostatic loads are not applicable and buoyant forces are either zero or nominal.
	Breakaway walls and other similar nonbearing elements must be designed and constructed to fail under design flood or lesser conditions (ASCE/SEI 24 Section 4.6.1). Flood openings are required in breakaway walls (ASCE/SEI 24 Section 4.6.2).	Lateral and vertical hydrostatic loads need not be considered for breakaway walls located below the DFE.

TABLE 2.6 Cases Where Hydrostatic Loads Do Not Need to Be Considered

upstream face of the building, drag forces along the sides of the building, and a negative force (suction) on the downstream face of the building. An approximation for the total hydrodynamic load, F_{dyn}, on a building or structural element can be determined by the following equation, which is applicable for all flow velocities, V (see Ref. 7 and Fig. 2.9 for a building with foundation walls and a building supported on piles):

$$F_{dyn} = \frac{1}{2} a \rho V^2 A \qquad (2.5)$$

This force is assumed to act at the stillwater mid-depth (that is, halfway between the SWEL and the eroded ground surface, GS; see Fig. 2.9) on the upstream face of the building or structural element. The terms in Eq. (2.5) are defined as follows:

- a = drag coefficient, which is a function of the shape of the object around which flow is directed. Recommended values of a are given in Fig. 2.9 (see ASCE/SEI 5.4.3 and Ref. 7).

- ρ = mass density of water. For fresh water, $\rho = 1.94$ slugs/ft^3 (1,000 kg/m^3) and for saltwater, $\rho = 1.99$ slugs/ft^3 (1,026 kg/m^3).

- V = design flood velocity, which is determined in accordance with Sec. 2.11 of this publication.

- A = surface area of a building, structure, or structural element normal to the water flow. For the foundation wall in Fig. 2.9, $A = wd_s$ where w is the width of the foundation wall perpendicular to the flow. For the pile, $A = Dd_s$ where D is the width of the rectangular pile perpendicular to the flow or the diameter of a round pile.

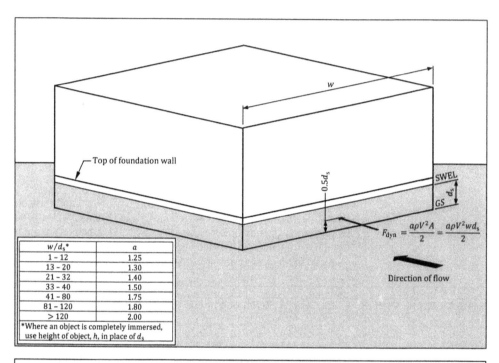

$$F_{dyn} = \frac{a\rho V^2 A}{2} = \frac{a\rho V^2 w d_s}{2}$$

w/d_s*	a
1 – 12	1.25
13 – 20	1.30
21 – 32	1.40
33 – 40	1.50
41 – 80	1.75
81 – 120	1.80
> 120	2.00

*Where an object is completely immersed, use height of object, h, in place of d_s

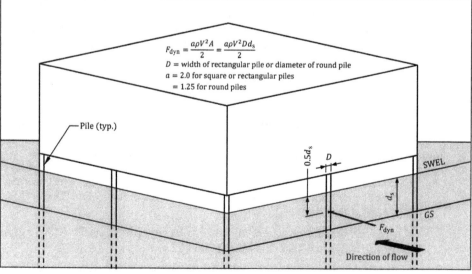

$$F_{dyn} = \frac{a\rho V^2 A}{2} = \frac{a\rho V^2 D d_s}{2}$$

D = width of rectangular pile or diameter of round pile
a = 2.0 for square or rectangular piles
  = 1.25 for round piles

Figure 2.9 Hydrodynamic loads on a foundation wall and on a pile.

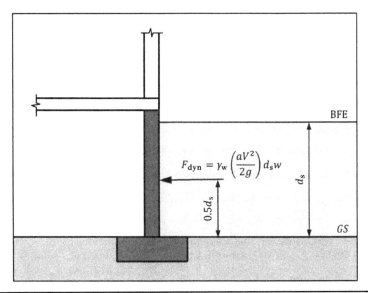

FIGURE 2.10 Hydrodynamic load where $V < 10$ ft/s (3.05 m/s).

The dynamic effects of moving water are permitted to be converted into an equivalent hydrostatic load where $V < 10$ ft/s (3.05 m/s) [ASCE/SEI 5.4.3]; this is accomplished by increasing the depth of the floodwater above the flood level by an amount d_h, which is determined by ASCE/SEI Equation (5.4-1), on the upstream face and above the ground level only:

$$d_h = \frac{aV^2}{2g} \tag{2.6}$$

The equivalent surcharge depth, d_h, is added to the DFE design depth and the resultant hydrostatic pressure is applied to and uniformly distributed across the vertical projected area of the building or structure perpendicular to the flow. The hydrodynamic load on the upstream face in such cases is equal to the following (see Fig. 2.10):

$$F_{dyn} = \gamma_w d_h d_s w = \gamma_w \left(\frac{aV^2}{2g}\right) d_s w = \frac{1}{2} a \left(\frac{\gamma_w}{g}\right) V^2 d_s w = \frac{1}{2} a \rho V^2 A \tag{2.7}$$

This load acts at $d_s/2$ from GS.

Surfaces of the building or structure other than the upstream face are subject to the hydrostatic pressures based on the depths to the BFE only.

2.14   Wave Loads

2.14.1   Overview

Wave loads result from water waves propagating over the surface of the water and striking a building or structural element. These types of loads are typically separated into four categories: nonbreaking waves, breaking waves, broken waves, and uplift (see Table 2.7).

Category	Comments
Nonbreaking waves	The effects of nonbreaking waves can be determined using the procedures in ASCE/SEI 5.4.2 for hydrostatic loads on walls and in ASCE/SEI 5.4.3 for hydrodynamic loads on piles.
Breaking waves	These loads are caused by waves breaking on any portion of a building or structure. Although these loads are of short duration, they generally produce the largest magnitude of all the different types of wave loads.
Broken waves	The loads caused by broken waves are similar to hydrodynamic loads caused by flowing or surging water.
Uplift	Uplift effects are caused by wave runup striking any portion of a building or structure, deflection, or peaking against the underside of surfaces.

TABLE 2.7   Categories of Wave Loads

Wave loads must be included in the design of buildings and structural elements located in (1) V Zones [wave heights greater than or equal to 3 ft (0.91 m)], (2) Coastal A Zones [wave heights greater than or equal to 1.5 ft (0.46 m) and less than 3 ft (0.91 m)], and (3) A Zones subject to coastal waves [wave heights less than 1.5 ft (0.46 m)].

Because breaking wave loads are the largest of the four types, it is used as the design wave load. Three methods are given in ASCE/SEI 5.4.4 to determine breaking wave loads:

- Analytical procedures
- Advanced numerical modeling procedures
- Laboratory test procedures (physical modeling)

Regardless of the method used, an estimate of the breaking wave height, H_b, is needed to calculate breaking waves loads (see Sec. 2.10 of this publication). Information on the FEMA model that can also be used to estimate wave heights and wave crest elevations can be found in Ref. 9.

A summary of the breaking wave loads and their points of application in accordance with the analytical procedures in ASCE/SEI 5.4.4 is given in Fig. 2.11.

2.14.2   Vertical Pilings and Columns

The net wave load, F_D, on vertical pilings and columns is applied to the element at the SWEL. The term C_D is the drag coefficient for breaking waves, which is equal to 1.75 for round piles or columns and 2.25 for square or rectangular piles or columns. The term D is the pile or column diameter for circular sections or 1.4 times the width of the pile or column for rectangular or square sections. The breaking wave height, H_b, is determined by Eq. (2.1).

Breaking wave loads must be considered, for example, on vertical pilings and columns supporting buildings in V Zones where the lowest floor in the building must be elevated to or above the minimum elevations identified in ASCE/SEI 24 Table 4-1 (see Sec. 2.8 of this publication).

2.14.3   Vertical Walls

For breaking wave loads on vertical walls, it is assumed in the analytical procedure the wall causes a reflected or standing wave to form against the waterward side of the wall and the crest of the reflected wave reaches a height of $1.2d_s$ above the SWEL. The resulting

Structural Element	Breaking Wave Load
Vertical Pilings and Columns	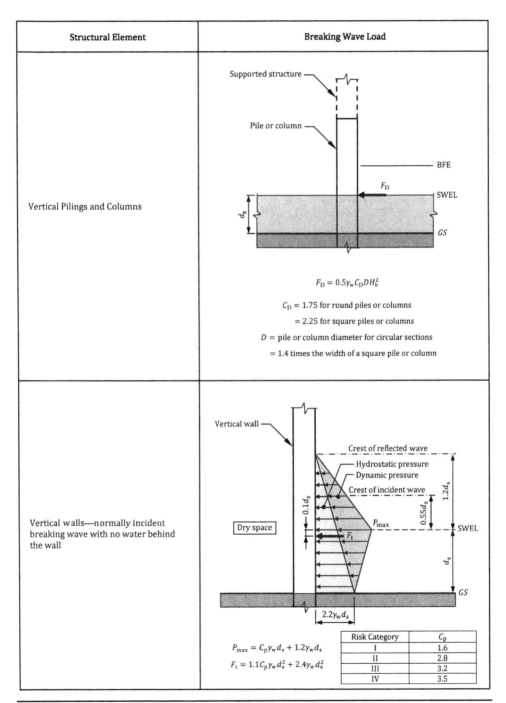
Vertical walls—normally incident breaking wave with no water behind the wall	

Vertical Pilings and Columns:

$$F_D = 0.5\gamma_w C_D D H_b^2$$

$C_D = 1.75$ for round piles or columns

$\quad = 2.25$ for square piles or columns

$D =$ pile or column diameter for circular sections

$\quad = 1.4$ times the width of a square pile or column

Vertical walls:

$$P_{max} = C_p \gamma_w d_s + 1.2\gamma_w d_s$$

$$F_t = 1.1 C_p \gamma_w d_s^2 + 2.4\gamma_w d_s^2$$

Risk Category	C_p
I	1.6
II	2.8
III	3.2
IV	3.5

Figure 2.11 Breaking wave loads in accordance with ASCE/SEI 5.4.4.

Structural Element	Breaking Wave Load
Vertical walls—normally incident breaking wave with stillwater level equal on both sides of wall	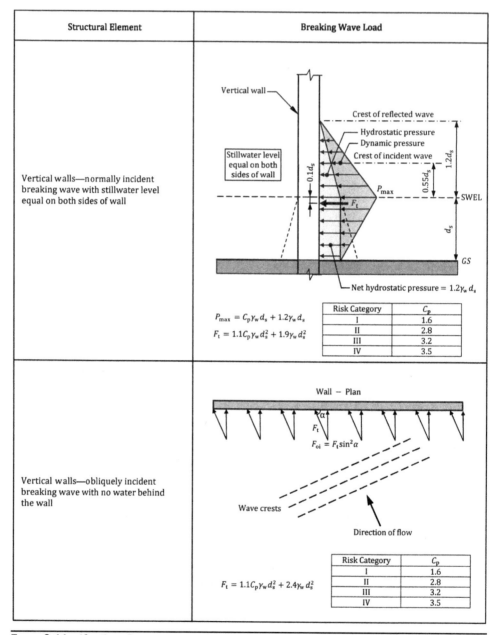

FIGURE 2.11 (Continued)

hydrostatic, dynamic, and total pressure distributions are given in Fig. 2.11 for the cases where there is no water behind the wall and where the stillwater level is equal on both sides of the wall.

The case of a vertical wall with no water behind it corresponds to nonresidential buildings in an A Zone where the lowest floor is below the minimum elevation in

Structural Element	Breaking Wave Load
Nonvertical walls—no water behind the wall	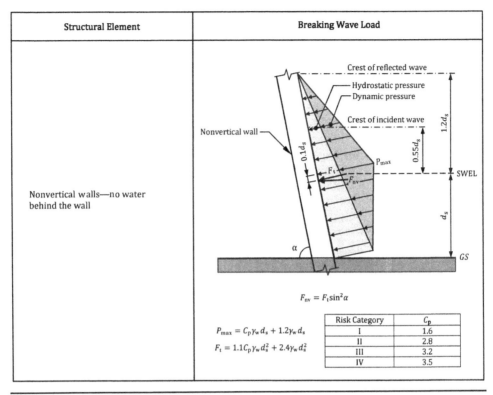

FIGURE **2.11** *(Continued)*

ASCE/SEI 24 Table 2-1 and the dry floodproofing requirements of ASCE/SEI 24 Section 6.2 are satisfied for the space behind the wall (see ASCE/SEI 24 Section 2.3; under the NFIP, solid foundation walls are not permitted in V Zones). In this case, the maximum total pressure, P_{max}, occurs at the SWEL and is equal to the maximum dynamic pressure $[C_p \gamma_w d_s]$ plus the hydrostatic pressure $[\gamma_w (1.2 d_s)]$ where C_p is the dynamic pressure coefficient, which is based on the risk category of the building (see ASCE/SEI Table 1.5-1 and ASCE/SEI Table 5.4-1). Risk Category II buildings are assigned a value of C_p corresponding to a 1 percent probability of exceedance, which is consistent with wave analysis procedures used by FEMA in mapping coastal FHAs and in establishing minimum floor elevations (see Sec. 2.8 of this publication). The percent probability of exceedance for the other risk categories is given in ASCE/SEI C5.4.4.2.

The net breaking wave load per unit length of wall (that is, the resultant force of the total pressure per unit length of wall), F_t, is equal to the resultant force of the dynamic pressure $[2.2 C_p \gamma_w d_s^2 (2.2 - 1.0)/(2 \times 1.2) = 1.1 C_p \gamma_w d_s^2]$ plus the resultant force of the hydrostatic pressure $[(2.2 \gamma_w d_s)(2.2 d_s)/2 = 2.4 \gamma_w d_s^2]$. It is assumed F_t is applied to the wall a distance equal to $0.1 d_s$ below the SWEL instead of at the centroid of the total pressure (Ref. 7).

Water can occur behind a wall where a wave breaks against (1) breakaway walls (which are required in V Zones and coastal A Zones for enclosed areas below the DFE; see ASCE/SEI 24 Section 4.6.1) or (2) walls equipped with openings that allow floodwaters to equalize on both sides of the wall. Based on the analytical procedure, the hydrostatic pressure equalized in this case is equal to $\gamma_w d_s$. Therefore, the net resultant force of the hydrostatic pressure over the height d_s of the wall is equal to $2.4\gamma_w d_s^2 - (\gamma_w d_s^2/2) = 1.9\gamma_w d_s^2$ (see Fig. 2.11). The resultant force of the dynamic pressure is equal to $1.1C_p\gamma_w d_s^2$, which is the same as that for a wall with no water behind it.

When designing breakaway walls, a value of 1.0 should be used for C_p rather than the values given in ASCE/SEI Table 5.4-1 (Ref. 7).

It is typically assumed the direction of wave approach is approximately perpendicular to the shoreline. For walls oriented perpendicular to the shoreline, the wave loads are maximum because the waves are normally incident to the wall; in such cases, the applicable equations in Fig. 2.11 can be used to calculate the breaking wave loads. For walls not oriented perpendicular to the shoreline, the waves are not normally incident to the wall and the corresponding loads are smaller than those for walls perpendicular to the shoreline. The breaking wave load from obliquely incident waves is $F_{oi} = F_t \sin^2\alpha$ where α is the horizontal angle between the direction of wave approach and the vertical surface of the wall [see ASCE/SEI Equation (5.4-9) and Fig. 2.11].

2.14.4   Nonvertical Walls

Breaking wave loads on nonvertical walls can be calculated by ASCE/SEI Equation (5.4-8): $F_{nv} = F_t \sin^2\alpha$ where α is the vertical angle between the nonvertical surface of the wall and the horizontal. The breaking wave pressures and loads illustrated in Fig. 2.11 are for the case where no water is behind the wall.

2.15   Impact Loads

2.15.1   Overview

Impact loads occur where objects carried by moving water strike a building or structural element. Although the magnitude of such loads is difficult to predict, reasonable approximations can be made considering a number of different uncertainties.

Impact loads can be categorized as normal, special, and extreme (Ref. 10). Information pertaining to these three categories is given in Table 2.8.

Impact Load Category	Comments
Normal	Normal impact loads result from the isolated impacts of commonly encountered objects. The size, shape, and weight of waterborne debris vary according to region.
Special	Special impact loads result from large objects such as broken up ice floats and accumulations of waterborne debris. The loads are caused by these objects either striking or resting against the building or structure.
Extreme	Extreme impact loads result from very large objects (such as boats, barges, or parts of collapsed buildings) striking a building or structure. Design for such extreme loads is usually not practical in most cases unless the probability of such an impact load during the design flood is high.

TABLE 2.8   Impact Loads

2.15.2 Normal Impact Loads

A rational method for calculating normal impact loads is given in ASCE/SEI C5.4.5. The flowchart in Fig. 2.12 can be used to determine the impact load, F [see ASCE/SEI Equation (C5.4-3)].

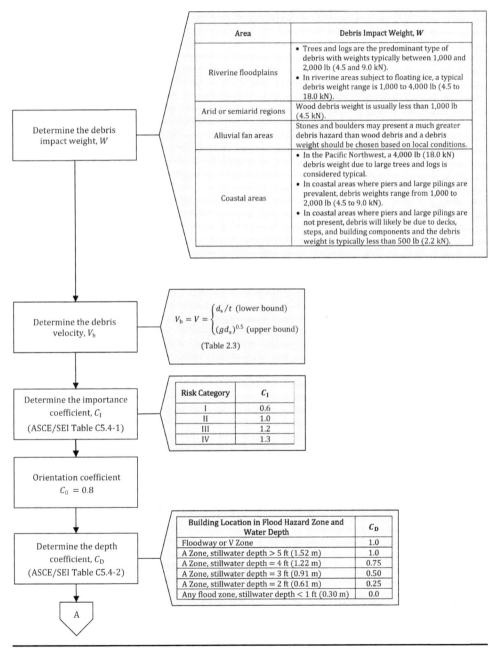

FIGURE 2.12 Flowchart to determine normal impact loads, F.

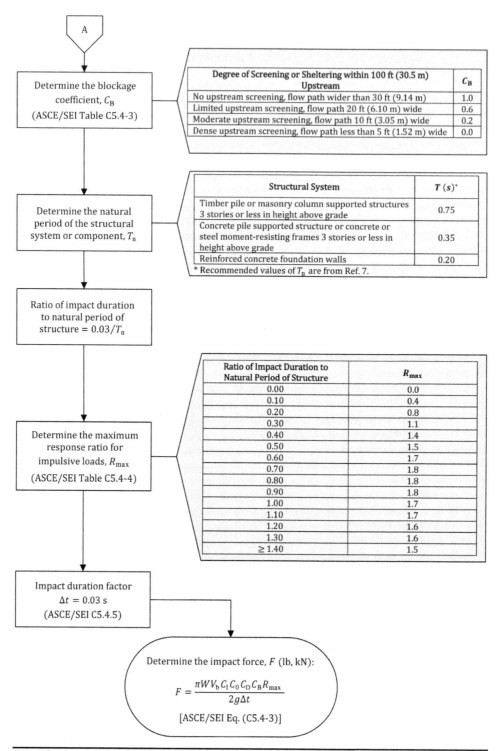

The figure (flowchart) contains the following elements:

A

Determine the blockage coefficient, C_B (ASCE/SEI Table C5.4-3)

Degree of Screening or Sheltering within 100 ft (30.5 m) Upstream	C_B
No upstream screening, flow path wider than 30 ft (9.14 m)	1.0
Limited upstream screening, flow path 20 ft (6.10 m) wide	0.6
Moderate upstream screening, flow path 10 ft (3.05 m) wide	0.2
Dense upstream screening, flow path less than 5 ft (1.52 m) wide	0.0

Determine the natural period of the structural system or component, T_n

Structural System	T $(s)^*$
Timber pile or masonry column supported structures 3 stories or less in height above grade	0.75
Concrete pile supported structure or concrete or steel moment-resisting frames 3 stories or less in height above grade	0.35
Reinforced concrete foundation walls	0.20

* Recommended values of T_n are from Ref. 7.

Ratio of impact duration to natural period of structure $= 0.03/T_n$

Determine the maximum response ratio for impulsive loads, R_{max} (ASCE/SEI Table C5.4-4)

Ratio of Impact Duration to Natural Period of Structure	R_{max}
0.00	0.0
0.10	0.4
0.20	0.8
0.30	1.1
0.40	1.4
0.50	1.5
0.60	1.7
0.70	1.8
0.80	1.8
0.90	1.8
1.00	1.7
1.10	1.7
1.20	1.6
1.30	1.6
≥ 1.40	1.5

Impact duration factor $\Delta t = 0.03$ s (ASCE/SEI C5.4.5)

Determine the impact force, F (lb, kN):

$$F = \frac{\pi W V_b C_I C_0 C_D C_B R_{max}}{2g\Delta t}$$

[ASCE/SEI Eq. (C5.4-3)]

FIGURE 2.12 *(Continued)*

Information on the terms used to calculate F is given in Table 2.9. The term $\pi/2$ in ASCE/SEI Equation (C5.4-3) is a result of the half-sine form of the impulse load used in the derivation of this equation. It is presumed objects are at or near the water surface level when they strike a building. Thus, F is usually assumed to act at the stillwater flood level; in general, this load should be applied horizontally at the most critical location at or below the SWEL.

Term	ASCE/SEI Reference	Comments
Debris weight, W	Section C5.4.5	The debris impact weight is to be selected considering local or regional conditions. A realistic weight for trees, logs, and other large wood-type debris, which are the most common forms of damaging debris, is 1,000 lb (4.5 kN). This is also a reasonable weight for small ice floes, boulders, and some man-made objects.
Velocity of object, V_b	Section C5.4.5	Debris velocity depends on the nature of the debris and the velocity of the floodwaters. Smaller pieces of debris usually travel at the velocity of the floodwaters; such items typically do not cause damage to buildings or structures. Large debris, which can cause damage, will likely travel at a velocity less than that of the floodwaters because it can drag at the bottom or can be slowed by prior collisions with nearby objects. For calculating debris loads, the velocity of the debris should be taken equal to the velocity of the floodwaters, V (see Sec. 2.11 of this publication).
Importance coefficient, C_I	Table C5.4-1	Importance coefficients are based on a probability distribution of impact loads obtained from laboratory tests and the magnitude of this coefficient is larger for more important structures.
Orientation coefficient, C_O	Section C5.4.5	The orientation coefficient is used to reduce the impact load for other than head-on impact loads. Based on measurements taken during laboratory tests, an orientation factor of 0.8 has been adopted.
Depth coefficient, C_D	Table C5.4-2 Figure C5.4-1	The depth coefficient is used to account for reduced debris velocity due to debris dragging along the bottom in shallow water. Values of the coefficient are based on typical diameters of logs and trees and on anticipated diameters of root masses from drifting trees likely to be encountered in a flood hazard Zone.
Blockage coefficient, C_B	Table C5.4-3 Figure C5.4-2	The blockage coefficient accounts for reduction of debris velocity due to screening and sheltering provided by trees or other structures within about 300 ft (91.4 m) upstream from a building or structure.

TABLE 2.9 Terms Used to Calculate Normal Impact Loads, F

Term	ASCE/SEI Reference	Comments
Maximum response ratio for impulsive loads, R_{max}	Table C5.4-4	The maximum response ratio modifies the impact load based on the natural period of the structure, T_n, and the impact duration, Δt. Stiff or rigid buildings and structural elements with natural periods similar to the impact duration will see an amplification of the impact load while more flexible buildings and structural elements with natural periods greater than approximately 4 times the impact duration will see a reduction of the impact load. The critical period is approximately 0.11 s: Buildings and structures with a natural period greater than 0.11 s will see a reduction in the impact load while those with a natural period less than 0.11 s will see an increase in the impact load. The natural period can be determined by any substantiated analysis. Recommendations on values of T_n to use for flood impact loads are given in ASCE/SEI C5.4.5 and in Ref. 7.
Impact duration time, Δt	Section C5.4.5	Impact duration is the time it takes to reduce the velocity of the object to zero. The recommended value of Δt is 0.03 s.

TABLE 2.9 Terms Used to Calculate Normal Impact Loads, F (*Continued*)

2.15.3 Special Impact Loads

The following equation can be used to determine special impact loads [see ASCE/SEI Equation (C5.4-4)]:

$$F = \frac{C_D \rho A V^2}{2} \tag{2.8}$$

In this equation, C_D is the drag coefficient, which is equal to 1.0; ρ is the mass density of water; A is the projected area of the debris accumulation into the flow, which is approximately the depth of the accumulation times the width of the accumulation perpendicular to the direction of the flow; and V is the velocity of the flow upstream of the debris accumulation. Like normal impact loads, this load is assumed to act at the still-water flood level unless it is found that the critical location is below the SWEL.

2.16 Flood Load Combinations

2.16.1 Flood Loads to Use in the Determination of F_a

A summary of the horizontal flood loads that need to be considered in the determination of the design flood load, F_a, for buildings located in an A Zone where the flood source is from a river or lake is given in Table 2.10. A similar summary is given in

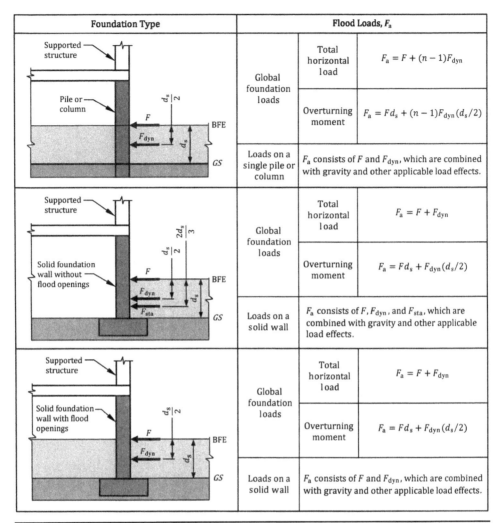

Foundation Type	Flood Loads, F_a		
	Global foundation loads	Total horizontal load	$F_a = F + (n-1)F_{dyn}$
		Overturning moment	$F_a = Fd_s + (n-1)F_{dyn}(d_s/2)$
	Loads on a single pile or column	F_a consists of F and F_{dyn}, which are combined with gravity and other applicable load effects.	
	Global foundation loads	Total horizontal load	$F_a = F + F_{dyn}$
		Overturning moment	$F_a = Fd_s + F_{dyn}(d_s/2)$
	Loads on a solid wall	F_a consists of F, F_{dyn}, and F_{sta}, which are combined with gravity and other applicable load effects.	
	Global foundation loads	Total horizontal load	$F_a = F + F_{dyn}$
		Overturning moment	$F_a = Fd_s + F_{dyn}(d_s/2)$
	Loads on a solid wall	F_a consists of F and F_{dyn}, which are combined with gravity and other applicable load effects.	

TABLE 2.10 Determination of F_a for Buildings in A Zones (Riverine or Lake Flood Source)

Table 2.11 for buildings located in A Zones subject to coastal waves, Coastal A Zones, and V Zones. The effects of these loads are used in the load combinations given in Sec. 2.16.2 of this publication.

A Zones—Riverine or Lake Flood Source (Not Subjected to Coastal Wave Loads)

- Buildings Supported by Piles or Columns

    In an A Zone where the source of the floodwater is from a river or lake and where the building is not subjected to wave loads, it is reasonable to assume for the analysis of the entire (global) foundation of a building or structure supported by n piles or columns that one of the piles or columns will be subjected to the impact load, F, and the remaining piles or columns will be subjected to the

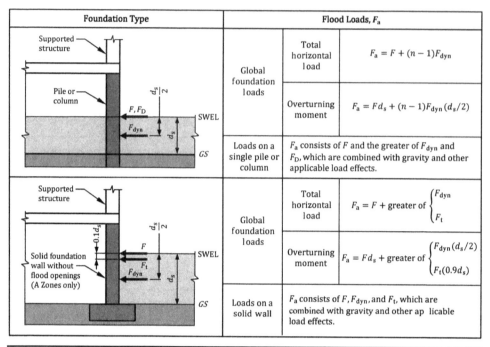

Foundation Type	Flood Loads, F_a		
	Global foundation loads	Total horizontal load	$F_a = F + (n-1)F_{dyn}$
		Overturning moment	$F_a = Fd_s + (n-1)F_{dyn}(d_s/2)$
	Loads on a single pile or column	F_a consists of F and the greater of F_{dyn} and F_D, which are combined with gravity and other applicable load effects.	
	Global foundation loads	Total horizontal load	$F_a = F + $ greater of $\begin{cases} F_{dyn} \\ F_t \end{cases}$
		Overturning moment	$F_a = Fd_s + $ greater of $\begin{cases} F_{dyn}(d_s/2) \\ F_t(0.9d_s) \end{cases}$
	Loads on a solid wall	F_a consists of F, F_{dyn}, and F_t, which are combined with gravity and other applicable load effects.	

TABLE 2.11 Determination of F_a for Buildings in A Zones (Coastal Waves), Coastal A Zones, and V Zones

hydrodynamic load, F_{dyn}. Therefore, the total horizontal load on the foundation due to flood loads in this case is equal to the following:

$$F_a = F + (n-1)F_{dyn} \qquad (2.9)$$

The total overturning moment about GS due to flood loads is equal to the following [see Table 2.10]:

$$F_a = Fd_s + (n-1)F_{dyn}(d_s/2) \qquad (2.10)$$

Individual piles or columns in the front row (that is, in the row experiencing the initial effects of the floodwaters) must be designed to resist the combined effects due to flood, gravity, and other applicable effects where F_a consists of F and F_{dyn}, which are applied to the pile or column at the locations indicated in Table 2.10.

- Buildings Supported by Solid Foundation Walls

  In the case of a solid foundation wall without flood openings where the dry floodproofing requirements of ASCE/SEI 24 Section 6.2 are satisfied, the total horizontal load on the foundation due to flood loads is equal to the following:

$$F_a = F + F_{dyn} \qquad (2.11)$$

The total overturning moment about GS is equal to the following:

$$F_a = Fd_s + F_{dyn}(d_s/2) \qquad (2.12)$$

When determining the global effects on the foundation in this case, it is evident the hydrostatic loads, F_{sta}, cancel out (see Table 2.10).

Individual foundation walls must be designed to resist the combined effects due to flood, gravity, and other applicable effects where F_a consists of F, F_{dyn}, and F_{sta}, which are applied to the wall at the locations indicated in Table 2.10. For foundation walls that extend below GS, F_{dif} must also be considered.

Equations (2.11) and (2.12) can also be used to determine global foundation effects on solid foundation walls with flood openings. Individual foundation walls in this case must be designed for the combined effects due to flood, gravity, and other applicable effects where F_a consists of F and F_{dyn}, which are applied to the wall at the locations indicated in Table 2.11. For foundation walls that extend below GS, F_{dif} must also be considered.

Breaking wave loads need not be considered in A Zones except at sites subject to coastal waves loads (see below).

The flood loads, F_a, determined above are used in the applicable load combinations given in Tables 2.12 through 2.15 of this publication. Load combinations that include vertical buoyant loads must also be considered where applicable.

A Zones (Subject to Coastal Wave Loads), Coastal A Zones, and V Zones

- Buildings Supported by Piles or Columns

  For a building or structure supported by n piles or columns, it is reasonable to assume that one of the piles or columns will be subjected to F and the remaining piles or columns will be subjected to F_{dyn}. Therefore, the total horizontal load on the foundation and the overturning moment about GS due to flood loads can be determined by Eqs. (2.9) and (2.10), respectively (see Table 2.11).

  Individual piles or columns in the front row must be designed to resist the combined effects due to flood, gravity, and other applicable effects where F_a consists of F and the greater of F_{dyn} and F_D, which are applied to the pile or column at the locations indicated in Table 2.11.

- Buildings Supported by Solid Foundation Walls

  As noted previously, solid foundation walls are not permitted by the IBC or ASCE/SEI 24 in Coastal A Zones or V Zones. For buildings in A Zones subjected to coastal wave loads, solid foundation walls without flood openings that meet the dry floodproofing requirements of ASCE/SEI 24 Section 6.2 are permitted, and the total horizontal load on the foundation due to flood loads in this case is equal to the following:

$$F_a = F + \text{greater of } \begin{cases} F_{dyn} \\ F_t \end{cases} \tag{2.13}$$

The total overturning moment about GS is equal to the following:

$$F_a = Fd_s + \text{greater of } \begin{cases} F_{dyn}(d_s/2) \\ F_t(0.9d_s) \end{cases} \tag{2.14}$$

The individual foundation walls must be designed to resist the combined effects due to flood, gravity, and other applicable effects where F_a consists of F and the greater of F_{dyn} and F_t, which are applied to the wall at the locations indicated in Table 2.11 (F_{sta} need not be considered in the design of the foundation wall because the hydrostatic component of the flood load is included in F_t). For foundation walls that extend below GS, F_{dif} must also be considered.

The flood loads, F_a, determined above are used in the applicable load combinations given in Tables 2.12 through 2.15 of this publication. Load combinations that include vertical buoyant loads must also be considered where applicable.

2.16.2   Strength Design and Allowable Stress Design Load Combinations

Strength design and allowable stress design load combinations pertaining to design flood loads, F_a, are given in ASCE/SEI 2.3.2 and 2.4.2, respectively. Strength design load combinations for structural elements above grade based on flood Zones are given in Table 2.12 and allowable stress design load combinations are given in Table 2.13. The load factors on F_a are based on a statistical analysis of flood loads associated with hydrostatic pressures, hydrodynamic pressures due to steady overland flow, and hydrodynamic pressures due to waves, all of which are specified in ASCE/SEI 5.4. All other applicable load combinations in ASCE/SEI 2.3 and 2.4 must also be considered in the design of the structural elements.

Flood Zone	ASCE/SEI Equation No.	Load Combination
A	4	$1.2D + 0.5W + 1.0F_a + L + 0.5(L_r$ or S or R$)$
	5	$0.9D + 0.5W + 1.0F_a$
Coastal A and V	4	$1.2D + 1.0W + 2.0F_a + L + 0.5(L_r$ or S or R$)$
	5	$0.9D + 1.0W + 2.0F_a$

TABLE 2.12   Strength Design Load Combinations Including Flood Loads for Structural Elements above Grade

Flood Zone	ASCE/SEI Equation No.	Load Combination
A	5	$D + 0.6W + 0.75F_a$
	6	$D + 0.75L + 0.75(0.6W) + 0.75(L_r$ or S or R$) + 0.75F_a$
	7	$0.6D + 0.6W + 0.75F_a$
Coastal A and V	5	$D + 0.6W + 1.5F_a$
	6	$D + 0.75L + 0.75(0.6W) + 0.75(L_r$ or S or R$) + 1.5F_a$
	7	$0.6D + 0.6W + 1.5F_a$

TABLE 2.13   Allowable Stress Design Load Combinations Including Flood Loads for Structural Elements above Grade

ASCE/SEI Equation No.	Load Combination
2	$1.2D + 1.6(L + H) + 0.5(L_r$ or S or $R)$
3	$1.2D + 1.6(L_r$ or S or $R) + (L$ or $0.5W) + 1.6H$
4	$1.2D + 1.0W + L + 0.5(L_r$ or S or $R) + 1.6H$
5	$0.9D + 1.0W + 1.6H$
6	$1.2D + 1.0E + L + 0.2S + 1.6H$
7	$0.9D + 1.0E + 1.6H$

Table 2.14 Strength Design Load Combinations Including Flood Loads for Structural Elements below Grade

ASCE/SEI Equation No.	Load Combination
2	$D + L + H$
3	$D + (L_r$ or S or $R) + H$
4	$D + 0.75L + 0.75(L_r$ or S or $R) + H$
5	$D + 0.6W + H$
6	$D + 0.75L + 0.75(0.6W) + 0.75(L_r$ or S or $R) + H$
7	$0.6D + 0.6W + H$
8	$D + 0.7E + H$
9	$D + 0.75(0.7E) + 0.75L + 0.75S + H$
10	$0.6D + 0.7E + H$

Table 2.15 Allowable Stress Design Load Combinations Including Flood Loads for Structural Elements below Grade

Strength design load combinations for structural elements below grade subjected to flood-induced hydrostatic uplift or lateral hydrostatic loads, H, are given in Table 2.14. Because the variability in hydrostatic loads under flood conditions is small compared with the variability in hydrodynamic loads from overland flooding and from wave loads, a load factor of 1.6 on H is used in all flood Zones instead of a load factor of 2.0, the latter of which is applicable to above-grade structural elements.

Allowable stress design load combinations for structural elements below grade subjected to flood-induced hydrostatic uplift or lateral hydrostatic loads, H, are given in Table 2.15.

2.17   Examples

The following examples illustrate the determination of flood loads in various flood hazard Zones. The steps shown in Fig. 2.1 are used to determine F_a.

2.17.1   Example 2.1—Foundation Wall Supporting a Residential Building Located in an AE Zone, Source of Floodwaters from a River

Determine the flood loads, F_a, on the foundation wall and footing supporting the residential building in Fig. 2.13. The location of the building is noted on the FIRM in Fig. 2.14. Assume the following:

- Source of the floodwaters is from a river, which is fresh water.
- BFE = 209.0 ft NAVD (63.7 m NAVD) from the FIRM (see Fig. 2.14).
- Dry floodproofing requirements of ASCE/SEI 24 Section 6.2 are met and no flood openings are provided in the foundation walls.

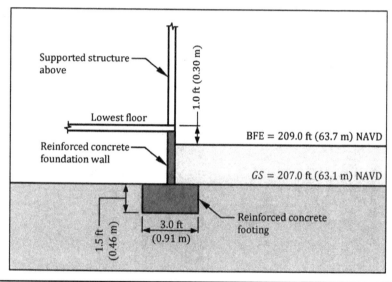

FIGURE 2.13   Residential building in Example 2.1.

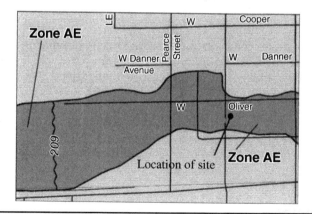

FIGURE 2.14   Location of the residential building in Example 2.1. (*Source: FEMA.*)

- The local jurisdiction has not adopted a more severe flood than that obtained from the FIRM.
- Floodwater debris hazards exist and are characterized as normal and special.
- The dimension of the building perpendicular to the floodwater flow is 40.0 ft (12.2 m).
- The thickness of the reinforced concrete foundation wall is 8 in. (203 mm).

Solution

Step 1—Acquire the flood hazard Zone

From the FIRM in Fig. 2.14, the building is located in an AE Zone. From the design data, the source of the floodwaters is from a river.

Step 2—Acquire the base flood elevation, BFE

From the FIRM in Fig. 2.14, BFE = 209.0 ft NAVD (63.7 m NAVD).

Step 3—Determine the design flood elevation, DFE

Because the local jurisdiction has not adopted a more severe flood than that obtained from the FIRM, DFE = BFE = 209.0 ft NAVD (63.7 m NAVD).

Step 4—Determine the minimum elevation of the lowest floor   Fig. 2.3

Because the building has a residential occupancy, the flood design class is 2 (see Table 1-1 in ASCE/SEI 24). Therefore, for an AE Zone, the minimum elevation of the top of the lowest floor is equal to the higher of the following:

- BFE + 1.0 = 209.0 + 1.0 = 210.0 ft NAVD (64.0 m NAVD)
- DFE = 209.0 ft NAVD (63.7 m NAVD)

Step 5—Determine the stillwater depth, d_s   Fig. 2.4

For riverine areas:

- SWEL = BFE = 209.0 ft (63.7 m)
- d_s = SWEL − GS = 209.0 − 207.0 = 2.0 ft (0.61 m)

Step 6—Determine the breaking wave height, H_b, where applicable   Fig. 2.5

Because the building is subjected to floodwaters from a river, wave loads are not applicable in this example and H_b need not be calculated.

Step 7—Determine the design flood velocity, V   Table 2.3

Because the building is located in an AE Zone, the lower bound average water velocity is appropriate:

$$V = d_s/t = 2.0/1.0 = 2.0 \text{ ft/s}$$

In S.I.:

$$V = d_s/t = 0.61/1.0 = 0.61 \text{ m/s}$$

Step 8—Determine the hydrostatic loads, F_{sta} and F_{buoy}, where applicable   Fig. 2.8

The lateral hydrostatic load, F_{sta}, per linear foot of foundation wall is determined by Eq. (2.3) with $w = 1.0$ ft:

$$F_{sta} = \gamma_w d_s^2 w/2$$

$$= 62.4 \times (2.0 + 1.0)^2 \times 1.0/2 = 281 \text{ lb/linear ft of foundation wall}$$

where d_s is increased by 1.0 ft in accordance with ASCE/SEI 5.4.2 because the foundation wall above ground is exposed to free water (Note: d_s need not be increased for other than this hydrostatic load).

This load acts at $3.0/3 = 1.0$ ft above GS.

In S.I.:

$$F_{sta} = \gamma_w d_s^2 / 2$$

$$= 9.80 \times (0.61 + 0.30)^2 \times 1.0/2 = 4.10 \text{ kN/linear m of foundation wall}$$

This load acts at $0.91/3 = 0.30$ m above GS.

There is no buoyancy load on the lowest floor slab.

The buoyancy load on the foundation wall is equal to the following per linear foot of foundation wall:

$$F_{buoy} = \gamma_w V_w$$

$$= \gamma_w (\text{wall thickness} \times d_s)$$

$$= 62.4 \times (8.0/12) \times 2.0 = 83 \text{ lb/linear ft of foundation wall}$$

In S.I.:

$$F_{buoy} = 9.80 \times (203/1,000) \times 0.61 = 1.21 \text{ kN/linear m of foundation wall}$$

The buoyancy load on the wall footing is equal to the following per linear foot of foundation wall:

$$F_{buoy} = \gamma_w V_f = 62.4 \times 1.5 \times 3.0 = 281 \text{ lb/linear ft of foundation wall}$$

In S.I.:

$$F_{buoy} = 9.80 \times 0.46 \times 0.91 = 4.10 \text{ kN/linear m of foundation wall}$$

Step 9—Determine the hydrodynamic load, F_{dyn}, where applicable   Fig. 2.9

$$F_{dyn} = \frac{1}{2} a \rho V^2 A \qquad \text{Eq. (2.5)}$$

For a building not totally immersed in floodwaters, use the ratio of the building width, w, to d_s to determine a:

For $w/d_s = 40.0/2.0 = 20.0$, $a = 1.30$.   Fig. 2.9

Therefore, F_{dyn} per linear foot of foundation wall is equal to the following:

$$F_{dyn} = \frac{a\rho V^2 w d_s}{2}$$

$$= \frac{1.30 \times 1.94 \times 2.0^2 \times 1.0 \times 2.0}{2} = 10 \text{ lb/linear ft of foundation wall}$$

This load acts at $d_s/2 = 2.0/2 = 1.0$ ft above GS.

Because $V < 10$ ft/s, F_{dyn} is permitted to be determined by Eq. (2.7) where the dynamic effects of moving water are converted into an equivalent hydrostatic load:

$$F_{dyn} = \gamma_w \left(\frac{aV^2}{2g}\right) d_s w = 62.4 \times \left(\frac{1.30 \times 2.0^2}{2 \times 32.2}\right) \times 2.0 \times 1.0$$

$$= 10 \text{ lb/linear ft of foundation wall}$$

In S.I.:

For $w/d_s = 12.2/0.61 = 20.0$, $a = 1.30$.   Fig. 2.9

Therefore, F_{dyn} per linear meter of foundation wall is equal to the following:

$$F_{dyn} = \frac{a\rho V^2 w d_s}{2} = \frac{1.30 \times 1.0 \times 0.61^2 \times 1.0 \times 0.61}{2}$$

$$= 0.15 \text{ kN/linear m of foundation wall}$$

This load acts at $d_s/2 = 0.61/2 = 0.30$ m above GS.

Because $V < 3.05$ m/s, F_{dyn} is permitted to be determined by Eq. (2.7) where the dynamic effects of moving water are converted into an equivalent hydrostatic load:

$$F_{dyn} = \gamma_w \left(\frac{aV^2}{2g}\right) d_s w = 9.80 \times \left(\frac{1.30 \times 0.61^2}{2 \times 9.81}\right) \times 0.61 \times 1.0$$

$$= 0.15 \text{ kN/linear m of foundation wall}$$

Step 10—Determine breaking wave loads, where applicable   Fig. 2.11

Breaking wave loads need not be determined in this example.

Step 11—Determine normal and special impact loads, F, where applicable

Normal impact loads are determined using the flowchart in Fig. 2.12.

- Determine the debris impact weight, W

  For riverine floodplains without floating ice, use an average:

  $$W = 1,500 \text{ lb (6.7 kN) [see the table in Fig. 2.12]}.$$

- Determine the debris velocity, V_b

  It is reasonable to assume $V_b = V = 2.0$ ft/s (0.61 m/s).

- Determine the importance coefficient, C_I

  For a Risk Category II building:

  $$C_I = 1.0 \qquad \text{ASCE/SEI Table C5.4-1}$$

- Orientation coefficient, $C_O = 0.8$. $\qquad$ ASCE/SEI C5.4.5
- Determine the depth coefficient, C_D

  For $d_s = 2.0$ ft (0.61 m) in an AE Zone:

$$C_D = 0.25 \qquad \text{ASCE/SEI Table C5.4-2}$$

- Determine the blockage coefficient, C_B

  Assuming there is no upstream screening and the flow path is wider than 30 ft (9.14 m):

$$C_B = 1.0 \qquad \text{ASCE/SEI Table C5.4-3}$$

- Determine the natural period, T_n, of the reinforced concrete foundation wall

  From the table in Fig. 2.12, approximate $T_n = 0.20$ s for reinforced concrete foundation walls.

- Determine the ratio of the impact duration to the natural period of structure, $\Delta t / T_n$

  Given the impact duration $\Delta t = 0.03$ s:

$$\Delta t / T_n = 0.03/0.20 = 0.15$$

- Determine the maximum response ratio for impulsive loads, R_{max}

  For $\Delta t / T_n = 0.15$:

$$R_{max} = 0.6 \text{ by linear interpolation} \qquad \text{ASCE/SEI Table C5.4-4}$$

- Determine the normal impact load, F $\qquad$ ASCE/SEI Equation (C5.4-3)

$$F = \frac{\pi W V_b C_I C_O C_D C_B R_{max}}{2g\Delta t} = \frac{\pi \times 1,500 \times 2.0 \times 1.0 \times 0.8 \times 0.25 \times 1.0 \times 0.6}{2 \times 32.2 \times 0.03}$$
$$= 585 \text{ lb}$$

In S.I.:

$$F = \frac{\pi W V_b C_I C_O C_D C_B R_{max}}{2g\Delta t} = \frac{\pi \times 6.7 \times 0.61 \times 1.0 \times 0.8 \times 0.25 \times 1.0 \times 0.6}{2 \times 9.81 \times 0.03}$$
$$= 2.6 \text{ kN}$$

This load acts at the SWEL = BFE, which is 2.0 ft (0.61 m) above the GS.

- Determine the special impact load, F

  Special impact loads are determined by Eq. (2.8):

$$F = \frac{C_D \rho A V^2}{2} = \frac{1.0 \times 1.94 \times (2.0 \times 40.0) \times 2.0^2}{2} = 310 \text{ lb}$$

In S.I.:

$$F = \frac{C_D \rho A V^2}{2} = \frac{1.0 \times 1.0 \times (0.61 \times 12.2) \times 0.61^2}{2} = 1.4 \text{ kN}$$

This load acts at the SWEL = BFE, which is 2.0 ft (0.61 m) above GS.

Step 12—Determine the flood loads to use in the determination of F_a   Table 2.10

- Global foundation loads

  For a solid foundation wall without openings that is dry floodproofed and located in an AE Zone where the source of the floodwaters is from a river, the total horizontal flood load on the foundation is equal to the following where the normal impact load is greater than the special impact load (see Table 2.10):

  $$F_a = F + F_{dyn} = 585 + (10 \times 40.0) = 985 \text{ lb}$$

  In S.I.:

  $$F_a = 2.6 + (0.15 \times 12.2) = 4.4 \text{ kN}$$

  The overturning moment about *GS* due to the flood loads is equal to the following:

  $$F_a = Fd_s + F_{dyn}(d_s/2) = (585 \times 2.0) + (10 \times 40.0 \times 2.0/2) = 1,570 \text{ ft-lb}$$

  In S.I.:

  $$F_a = (2.6 \times 0.61) + (0.15 \times 12.2 \times 0.61/2) = 2.14 \text{ kN-m}$$

- Foundation wall loads

  The flood loads on an individual foundation wall consist of F, F_{dyn}, and F_{sta} (see Fig. 2.15). The load effects due to these loads are combined with the other applicable load effects using the appropriate load combinations in Sec. 2.16.2 of this publication.

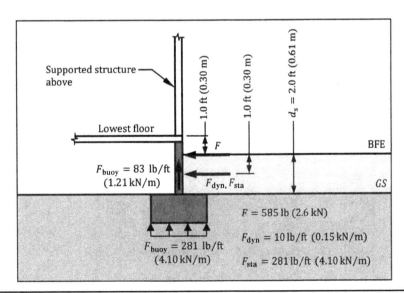

Figure 2.15 Flood loads for the residential building in Example 2.1.

2.17.2   Example 2.2—Foundation Wall Supporting a Nonresidential Building Located in an AE Zone, Source of Floodwaters from Coastal Waters

Determine the flood loads, F_a, on the foundation wall and footing supporting the nonresidential building in Fig. 2.16. The location of the building is noted on the FIRM in Fig. 2.17. Assume the following:

- Source of the floodwaters is from coastal waters, which is salt water.
- The ground slopes up gently from the shoreline and there are nominal obstructions between the shoreline and the site.
- The nonresidential building has a business occupancy and is oriented perpendicular to the shoreline.
- From the FIRM, the flood hazard Zone is AE and BFE = 5.0 ft NAVD (1.52 m NAVD).
- From the FIS, SWEL = 4.0 ft NAVD (1.22 m NAVD).
- GS = 3.0 ft NAVD (0.91 m NAVD)
- Dry floodproofing requirements of ASCE/SEI 24 Section 6.2 are met and no flood openings are provided in the foundation walls.
- The local jurisdiction requires a 1.0 ft (0.30 m) freeboard.

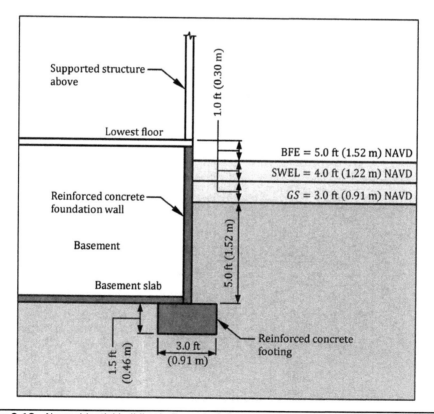

Figure 2.16   Nonresidential building in Example 2.2.

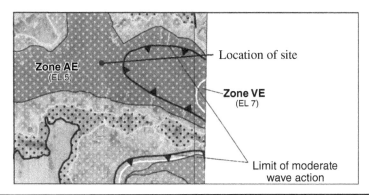

FIGURE 2.17   Location of the nonresidential building in Example 2.2. (*Source: FEMA.*)

- Floodwater debris hazards exist and are characterized as normal and special.
- The building plan dimensions are 80.0 ft by 80.0 ft (24.4 m by 24.4 m).
- The soil is clayey sand (SC) with $\gamma_s = 82$ lb/ft³ (12.88 kN/m³).
- The thickness of the foundation wall is 10.0 in. (254 mm).

Solution
Step 1—Acquire the flood hazard Zone

From the FIRM, the building is located in an AE Zone and the source of the floodwaters is from coastal waters.

The site is not located in a Coastal A Zone because it is landward of the LiMWA (see Fig. 2.17).

Step 2—Acquire the base flood elevation, BFE

From the FIRM, BFE = 5.0 ft NAVD (1.52 m NAVD).

Step 3—Determine the design flood elevation, DFE

Because the local jurisdiction has adopted a more severe flood than that obtained from the FIRM, DFE = BFE + freeboard = 5.0 + 1.0 = 6.0 ft NAVD (1.83 m NAVD).

Step 4—Determine the minimum elevation of the lowest floor   Fig. 2.3

Because the building has a business occupancy, the flood design class is 2 (see Table 1-1 in ASCE/SEI 24). Therefore, for an AE Zone, the minimum elevation of the top of the lowest floor is equal to the higher of the following:

- BFE + 1.0 = 5.0 + 1.0 = 6.0 ft NAVD (1.83 m NAVD)
- DFE = 6.0 ft NAVD (1.83 m NAVD).

The basement in this nonresidential building is permitted to be located below the minimum elevation in ASCE/SEI 24 Table 2-1 because the dry floodproofing requirements in ASCE/SEI 24 Section 6.2 are satisfied.

Step 5—Determine the stillwater depth, d_s

$$d_s = SWEL - GS = 4.0 - 3.0 = 1.0 \text{ ft (0.30 m)}$$

Step 6—Determine the breaking wave height, H_b, where applicable   Fig. 2.5

For a gently sloping shoreline with nominal obstructions:

$$H_b = 0.78\,d_s = 0.78 \times 1.0 = 0.8 \text{ ft (0.23 m)}$$ Eq. (2.1)

Step 7—Determine the design flood velocity, V   Table 2.3

Because the building is located in an AE Zone, the lower bound average water velocity is appropriate:

$$V = d_s/t = 1.0/1.0 = 1.0 \text{ ft/s}$$

In S.I.:

$$V = d_s/t = 0.30/1.0 = 0.30 \text{ m/s}$$

Step 8—Determine the hydrostatic loads, F_{sta}, F_{dif}, and F_{buoy}, where applicable   Fig. 2.8

The lateral hydrostatic load, F_{sta}, on the potion of the foundation wall above GS need not be determined because breaking wave loads must be considered, which include the hydrostatic component of the flood load (see Step 10).

Below GS, F_{sta} is equal to the following per linear foot of foundation wall with $w = 1.0$ ft:

$$F_{sta} = \gamma_w (\text{distance from GS to the top of the footing})^2 w/2$$

$$= 64.0 \times 5.0^2 \times 1.0/2 = 800 \text{ lb/ft}$$

This load acts at $5.0/3 = 1.7$ ft above the top of the footing.
In S.I.:

$$F_{sta} = 10.05 \times 1.52^2 \times 1.0/2 = 11.6 \text{ kN/m}$$

This load acts at $1.52/3 = 0.51$ m above the top of the footing.

The lateral load on the foundation wall below GS due to the differential between the water and soil pressures, F_{dif}, is determined by the following equation with $w = 1.0$ ft (see Fig. 2.8):

$$F_{dif} = (\gamma_s - \gamma_w)(\text{distance from GS to the top of the footing})^2 w/2$$

$$= (82.0 - 64.0) \times 5.0^2 \times 1.0/2 = 225 \text{ lb/linear ft of foundation wall}$$

This load acts at $5.0/3 = 1.7$ ft above the top of the footing.
In S.I.:

$$F_{dif} = (12.88 - 10.05) \times 1.52^2 \times 1.0/2 = 3.27 \text{ kN/linear m of foundation wall}$$

This load acts at $1.52/3 = 0.51$ m above the top of the footing.

The buoyancy pressure on the basement slab is determined by the following equation (see Fig. 2.8):

$$f_{buoy} = \gamma_w (\text{distance from the top of the flood water to the top of the footing})$$

$$= 64.0 \times (1.0 + 5.0) = 384 \text{ lb/ft}^2$$

With $A_b = 80.0 \times 80.0 = 6,400$ ft^2, the buoyancy load, F_{buoy}, is equal to the following:

$$F_{buoy} = 384 \times 6,400/1,000 = 2,458 \text{ kips}$$

In S.I.:

$$f_{buoy} = 10.05 \times (0.30 + 1.52) = 18.3 \text{ kN/m}^2$$

With $A_b = 24.4 \times 24.4 = 595.4$ m^2, the buoyancy load, F_{buoy}, is equal to the following:

$$F_{buoy} = 18.3 \times 595.4 = 10,896 \text{ kN}$$

The buoyancy load on the foundation wall is equal to the following per linear foot of foundation wall:

$$F_{buoy} = \gamma_w V_w = \gamma_w (\text{wall thickness} \times \text{submerged depth of the wall})$$

$$= 64.0 \times (10.0/12) \times (1.0 + 5.0) = 320 \text{ lb/linear ft of foundation wall}$$

In S.I.:

$$F_{buoy} = 10.05 \times (254/1,000) \times (0.30 + 1.52)$$
$$= 4.65 \text{ kN/linear m of foundation wall}$$

The buoyancy load on the wall footing is equal to the following per linear foot of foundation wall:

$$F_{buoy} = \gamma_w V_f = 64.0 \times 1.5 \times 3.0 = 288 \text{ lb/linear ft of foundation wall}$$

In S.I.:

$$F_{buoy} = 10.05 \times 0.46 \times 0.91 = 4.21 \text{ kN/linear m of foundation wall}$$

Step 9—Determine the hydrodynamic load, F_{dyn}, where applicable   Fig. 2.9

$$F_{dyn} = \frac{1}{2}a\rho V^2 A \qquad\qquad \text{Eq. (2.5)}$$

For a building not totally immersed in floodwaters, use the ratio of the building width, w, to d_s to determine a:

$$\text{For } w/d_s = 80.0/1.0 = 80.0, a = 1.75. \qquad\qquad \text{Fig. 2.9}$$

Therefore, F_{dyn} per linear foot of foundation wall is equal to the following:

$$F_{dyn} = \frac{a\rho V^2 w d_s}{2} = \frac{1.75 \times 1.99 \times 1.0^2 \times 1.0 \times 1.0}{2}$$
$$= 1.7 \text{ lb/linear ft of foundation wall}$$

This load acts at $d_s/2 = 1.0/2 = 0.5$ ft above *GS*.

In S.I.:

$$F_{dyn} = \frac{a\rho V^2 w d_s}{2} = \frac{1.75 \times 1.026 \times 0.30^2 \times 1.0 \times 0.30}{2}$$

$$= 0.024 \text{ kN/linear m of foundation wall}$$

This load acts at $d_s/2 = 0.30/2 = 0.15$ m above GS.

Step 10—Determine breaking wave loads, where applicable   Fig. 2.11

Breaking wave loads occur on the vertical foundation wall. Because flood openings are not provided in the foundation wall to allow passage of the floodwaters, the breaking wave load, F_t, is determined by ASCE/SEI Equation (5.4-6), which is applicable where the space behind the vertical wall is dry:

$$F_t = 1.1 C_p \gamma_w d_s^2 + 2.4 \gamma_w d_s^2$$

For a Risk Category II building, $C_p = 2.8$.   ASCE/SEI Table 5.4-1

Therefore, F_t per linear foot of foundation wall is equal to the following:

$$F_t = (1.1 \times 2.8 \times 64.0 \times 1.0^2) + (2.4 \times 64.0 \times 1.0^2)$$

$$= 351 \text{ lb/linear ft of foundation wall}$$

This load acts at $0.1 d_s = 0.1$ ft below the SWEL (for practical purposes, assume F_t acts at the SWEL, which is 1.0 ft above GS).
    In S.I.:

$$F_t = (1.1 \times 2.8 \times 10.05 \times 0.30^2) + (2.4 \times 10.05 \times 0.30^2)$$

$$= 4.96 \text{ kN/linear m of foundation wall}$$

This load acts at $0.1 d_s = 0.03$ m below the SWEL (for practical purposes, assume F_t acts at the SWEL, which is 2.30 m above GS).

Step 11—Determine normal and special impact loads, F, where applicable

Normal impact loads are determined using the flowchart in Fig. 2.12.

- Determine the debris impact weight, W

  For coastal areas where piers and large pilings are not present, use $W = 500$ lb (2.3 kN) [see the table in Fig. 2.12].

- Determine the debris velocity, V_b

  It is reasonable to assume $V_b = V = 1.0$ ft/s (0.30 m/s).

- Determine the importance coefficient, C_I

  For a Risk Category II building:

$$C_I = 1.0$$ ASCE/SEI Table C5.4-1

- Orientation coefficient, $C_O = 0.8$.   ASCE/SEI C5.4.5

- Determine the depth coefficient, C_D

  A value of C_D is not given in ASCE/SEI Table C5.4-2 for $d_s = 1.0$ ft (0.30 m) in an AE Zone.

  $$\text{Use } C_D = 0.25.$$

- Determine the blockage coefficient, C_B

  Assuming there is no upstream screening and the flow path is wider than 30 ft (9.14 m):

  $$C_B = 1.0 \qquad\qquad \text{ASCE/SEI Table C5.4-3}$$

- Determine the natural period, approximate T_n, of the reinforced concrete foundation wall.

  From the table in Fig. 2.12, $T_n = 0.20$ s for reinforced concrete foundation walls.

- Determine the ratio of the impact duration to the natural period of structure, $\Delta t / T_n$

  Given the impact duration $\Delta t = 0.03$ s:

  $$\Delta t / T_n = 0.03/0.20 = 0.15.$$

- Determine the maximum response ratio for impulsive loads, R_{max}

  For $\Delta t / T_n = 0.15$:

  $$R_{max} = 0.6 \text{ by linear interpolation} \qquad \text{ASCE/SEI Table C5.4-4}$$

- Determine the normal impact force, F $\qquad\qquad$ ASCE/SEI Equation (C5.4-3)

$$F = \frac{\pi W V_b C_I C_O C_D C_B R_{max}}{2 g \Delta t} = \frac{\pi \times 500 \times 1.0 \times 1.0 \times 0.8 \times 0.25 \times 1.0 \times 0.6}{2 \times 32.2 \times 0.03} = 98 \text{ lb}$$

In S.I.:

$$F = \frac{\pi W V_b C_I C_O C_D C_B R_{max}}{2 g \Delta t} = \frac{\pi \times 2.2 \times 0.30 \times 1.0 \times 0.8 \times 0.25 \times 1.0 \times 0.6}{2 \times 9.81 \times 0.03}$$
$$= 0.42 \text{ kN}$$

This load acts at the SWEL, which is 1.0 ft (0.30 m) above GS. Special impact loads are determined by Eq. (2.8):

$$F = \frac{C_D \rho A V^2}{2} = \frac{1.0 \times 1.99 \times (1.0 \times 80.0) \times 1.0^2}{2} = 80 \text{ lb}$$

In S.I.:

$$F = \frac{C_D \rho A V^2}{2} = \frac{1.0 \times 1.026 \times (0.30 \times 24.4) \times 0.30^2}{2} = 0.34 \text{ kN}$$

This load acts at the SWEL, which is 1.0 ft (0.30 m) above GS.

Step 12—Determine the flood loads to use in the determination of F_a   Table 2.11

- Global foundation loads

For a solid foundation wall without openings that is dry floodproofed and located in an AE Zone where the source of the floodwaters is from coastal waters, the total horizontal flood load on the foundation is equal to the following where the normal impact load is greater than the special impact load (see Table 2.11):

$$F_a = \text{greater of} \begin{cases} F + F_{dyn} = 98 + (1.7 \times 80.0) = 234 \text{ lb} \\ \\ F + F_t = 98 + (351 \times 80.0) = 28,178 \text{ lb} \end{cases}$$

In S.I.:

$$F_a = \text{greater of} \begin{cases} F + F_{dyn} = 0.42 + (0.024 \times 24.4) = 1.0 \text{ kN} \\ \\ F + F_t = 0.42 + (4.96 \times 24.4) = 121.4 \text{ kN} \end{cases}$$

The overturning moment about *GS* due to the flood loads is equal to the following assuming F_t acts at the SWEL:

$$F_a = \text{greater of} \begin{cases} Fd_s + F_{dyn}(d_s/2) = (98 \times 1.0) + (1.7 \times 80.0 \times 1.0/2) = 166 \text{ ft-lb} \\ \\ Fd_s + F_t d_s = (98 \times 1.0) + (351 \times 80.0 \times 1.0) = 28,178 \text{ ft-lb} \end{cases}$$

In S.I.:

$$F_a = \text{greater of} \begin{cases} (0.42 \times 0.30) + (0.024 \times 24.4 \times 0.30/2) = 0.21 \text{ kN-m} \\ \\ (0.42 \times 0.30) + (4.96 \times 24.4 \times 0.30) = 36.4 \text{ kN-m} \end{cases}$$

- Foundation wall loads

The flood loads on an individual foundation wall consist of F, F_{dyn}, F_t, F_{sta}, and F_{dif} (see Fig. 2.18). The load effects due to these loads are combined with the other applicable load effects using the appropriate load combinations in Sec. 2.16.2 of this publication.

2.17.3   Example 2.3—Columns Supporting a Nonresidential Building Located in a Coastal A Zone

Determine the flood loads, F_a, on the reinforced concrete columns and the reinforced concrete mat foundation supporting the nonresidential building depicted in Fig. 2.19. The location of the building is noted on the FIRM in Fig. 2.20. Assume the following:

- Source of the floodwaters is from coastal waters, which is salt water.
- The ground slopes up gently from the shoreline and there are nominal obstructions between the shoreline and the site.

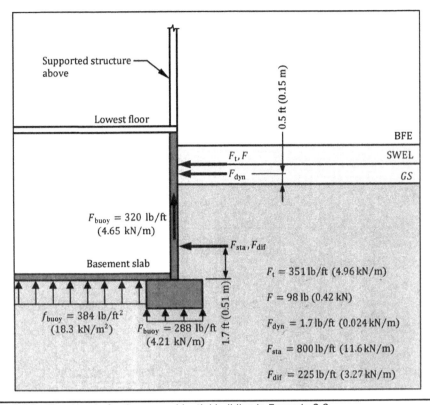

Figure 2.18   Flood loads for the nonresidential building in Example 2.2.

- The nonresidential building has a business occupancy and is oriented perpendicular to the shoreline.
- From the FIRM, the flood hazard Zone is AE and BFE = 5.0 ft NAVD (1.52 m NAVD).
- From the FIS, SWEL = 3.0 ft NAVD (0.91 m NAVD).
- GS = 0.0 ft NAVD (0.0 m NAVD).
- The local jurisdiction requires a 1.0 ft (0.30 m) freeboard.
- Floodwater debris hazards exist and are characterized as normal and special.

Solution

Step 1—Acquire the flood hazard Zone

From the design data, the building is located in an AE Zone and the source of the floodwaters is from coastal waters. Because the site is landward of the VE Zone and seaward of the LiMWA, it is designated a Coastal A Zone (see Fig. 2.20).

Step 2—Acquire the base flood elevation, BFE

From the FIRM, BFE = 5.0 ft NAVD (1.52 m NAVD).

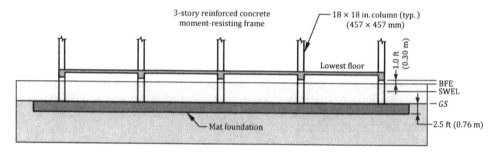

3-story reinforced concrete moment-resisting frame

18 × 18 in. column (typ.) (457 × 457 mm)

Lowest floor

1.0 ft (0.30 m)

BFE
SWEL
GS

Mat foundation

2.5 ft (0.76 m)

BFE = 5.0 ft NAVD (1.52 m NAVD)
SWEL = 3.0 ft NAVD (0.91 m NAVD)
GS = 0.0 ft NAVD (0.0 m NAVD)

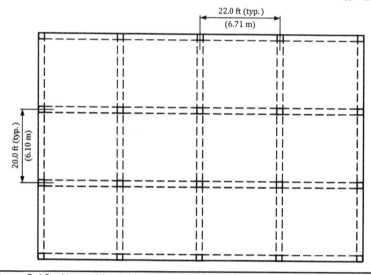

22.0 ft (typ.)
(6.71 m)

20.0 ft (typ.)
(6.10 m)

Figure 2.19 Nonresidential building in Example 2.3.

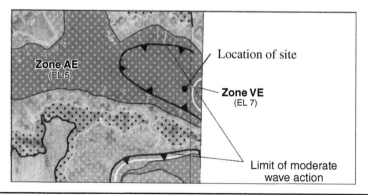

Zone AE (EL 5)

Location of site

Zone VE (EL 7)

Limit of moderate wave action

Figure 2.20 Location of the nonresidential building in Example 2.3.

Step 3—Determine the design flood elevation, DFE

Because the local jurisdiction has adopted a more severe flood than that obtained from the FIRM, DFE = BFE + freeboard = 5.0 + 1.0 = 6.0 ft NAVD (1.83 m NAVD).

Step 4—Determine the minimum elevation of the lowest floor   Fig. 2.3

Because the building has a business occupancy, the flood design class is 2 (see Table 1-1 in ASCE/SEI 24). Therefore, for a Coastal A Zone, the minimum elevation of the bottom of the lowest supporting horizontal structural member is equal to the higher of the following:

- BFE + 1.0 = 5.0 + 1.0 = 6.0 ft NAVD (1.83 m NAVD)

- DFE = 6.0 ft NAVD (1.83 m NAVD).

Step 5—Determine the stillwater depth, d_s

$$d_s = \text{SWEL} - \text{GS} = 3.0 - 0.0 = 3.0 \text{ ft } (0.91 \text{ m})$$

Step 6—Determine the breaking wave height, H_b, where applicable   Fig. 2.5

For a gently sloping shoreline with nominal obstructions:

$$H_b = 0.78 d_s = 0.78 \times 3.0 = 2.3 \text{ ft } (0.71 \text{ m})$$ Eq. (2.1)

Step 7—Determine the design flood velocity, V   Table 2.3

Because the building is located in a Coastal A Zone, the upper bound average water velocity is appropriate:

$$V = (g d_s)^{0.5} = (32.2 \times 3.0)^{0.5} = 9.8 \text{ ft/s}$$

In S.I.:

$$V = (g d_s)^{0.5} = (9.81 \times 0.91)^{0.5} = 2.99 \text{ m/s}$$

Step 8—Determine the hydrostatic loads, F_{sta} and F_{buoy}, where applicable   Fig. 2.8

Lateral hydrostatic loads on the columns are included in the hydrodynamic load, F_{dyn}.

There is no buoyancy load on the lowest floor slab. The buoyancy pressure on the mat foundation is equal to the following:

$$f_{buoy} = \gamma_w (\text{thickness of the mat foundation}) = 64.0 \times 2.5 = 160 \text{ lb/ft}^2$$

In S.I.:

$$f_{buoy} = 10.05 \times 0.76 = 7.6 \text{ kN/m}^2$$

This buoyancy pressure is uniformly distributed over the entire base area of the mat foundation.

The buoyancy load on a column is equal to the following:

$$F_{buoy} = \gamma_w d_s A_c = 64.0 \times 3.0 \times (18.0/12)^2 = 432 \text{ lb}$$

In S.I.:

$$F_{buoy} = \gamma_w d_s A_c = 10.05 \times 0.91 \times (457/1{,}000)^2 = 1.9 \text{ kN}$$

Step 9—Determine the hydrodynamic load, F_{dyn}, where applicable   Fig. 2.9

Hydrodynamic loads occur on the columns and are determined by the following equation:

$$F_{dyn} = \frac{1}{2}a\rho V^2 A$$ Eq. (2.5)

From Fig. 2.9, use $a = 2.0$ for rectangular columns.

Therefore, F_{dyn} on one of the columns is equal to the following:

$$F_{dyn} = \frac{a\rho V^2 Dd_s}{2} = \frac{2.0\times1.99\times9.8^2\times(18.0/12)\times3.0}{2} = 860 \text{ lb}$$

This load acts at $d_s/2 = 3.0/2 = 1.5$ ft above GS.

In S.I.:

$$F_{dyn} = \frac{a\rho V^2 Dd_s}{2} = \frac{2.0\times1.026\times2.99^2\times(457/1,000)\times0.91}{2} = 3.8 \text{ kN}$$

This load acts at $d_s/2 = 0.91/2 = 0.46$ m above GS.

Step 10—Determine breaking wave loads, where applicable   Fig. 2.11

Breaking wave loads, F_D, occur on the vertical columns and are determined by ASCE/SEI Equation (5.4-4):

$$F_D = 0.5\gamma_w C_D DH_b^2$$

$C_D = 2.25$ for square columns.   ASCE/SEI 5.4.4.1

$D = 1.4\times(18.0/12) = 2.1$ ft   ASCE/SEI 5.4.4.1

Therefore, F_D on one of the columns is equal to the following:

$$F_D = 0.5\times64.0\times2.25\times2.1\times2.3^2 = 800 \text{ lb}$$

This load acts at the SWEL, which is 3.0 ft above GS.

In S.I.:

$$D = 1.4\times(457/1,000) = 0.64 \text{ m}$$

$$F_D = 0.5\times10.05\times2.25\times0.64\times0.71^2 = 3.7 \text{ kN}$$

This load acts at the SWEL, which is 0.91 m above GS.

Step 11—Determine normal and special impact loads, F, where applicable

Normal impact loads are determined using the flowchart in Fig. 2.12.

- Determine the debris impact weight, W

  For coastal areas where piers and large pilings are present, use $W = 2,000$ lb (9.0 kN) [see the table in Fig. 2.12].

- Determine the debris velocity, V_b

  It is reasonable to assume $V_b = V = 9.8$ ft/s (2.99 m/s).

- Determine the importance coefficient, C_I

  For a Risk Category II building:

  $$C_I = 1.0 \qquad \text{ASCE/SEI Table C5.4-1}$$

- Orientation coefficient, $C_O = 0.8$. \hfill ASCE/SEI C5.4.5

- Determine the depth coefficient, C_D

  For $d_s = 3.0$ ft (0.91 m) in a Coastal A Zone:

  $$C_D = 0.50$$

- Determine the blockage coefficient, C_B

  Assuming there is no upstream screening and the flow path is wider than 30 ft (9.14 m):

  $$C_B = 1.0 \qquad \text{ASCE/SEI Table C5.4-3}$$

- Determine the natural period, T_n, of the reinforced concrete moment-resisting frame.

  From the table in Fig. 2.12, approximate $T_n = 0.35$ s for a reinforced concrete moment-resisting frame 3 stories in height.

- Determine the ratio of the impact duration to the natural period of structure, $\Delta t / T_n$.

  Given the impact duration $\Delta t = 0.03$ s:

  $$\Delta t / T_n = 0.03/0.35 = 0.09$$

- Determine the maximum response ratio for impulsive loads, R_{max}

  For $\Delta t / T_n = 0.09$:

  $$R_{max} = 0.4 \qquad \text{ASCE/SEI Table C5.4-4}$$

- Determine the normal impact force, F \hfill ASCE/SEI Equation (C5.4-3)

$$F = \frac{\pi W V_b C_I C_O C_D C_B R_{max}}{2g\Delta t}$$

$$= \frac{\pi \times 2,000 \times 9.8 \times 1.0 \times 0.8 \times 0.50 \times 1.0 \times 0.4}{2 \times 32.2 \times 0.03} = 5,099 \text{ lb}$$

In S.I.:

$$F = \frac{\pi W V_b C_I C_O C_D C_B R_{max}}{2g\Delta t}$$

$$= \frac{\pi \times 9.0 \times 2.99 \times 1.0 \times 0.8 \times 0.50 \times 1.0 \times 0.4}{2 \times 9.81 \times 0.03} = 23.0 \text{ kN}$$

This load acts at the SWEL, which is 3.0 ft (0.91 m) above *GS*.

Special impact loads are determined by Eq. (2.8):

$$F = \frac{C_D \rho A V^2}{2} = \frac{1.0 \times 1.99 \times [(18.0/12) \times 3.0] \times 9.8^2}{2} = 430 \text{ lb}$$

In S.I.:

$$F = \frac{C_D \rho A V^2}{2} = \frac{1.0 \times 1.026 \times (0.457 \times 0.91) \times 2.99^2}{2} = 1.9 \text{ kN}$$

This load acts at the SWEL, which is 3.0 ft (0.91 m) above *GS*.

Step 12—Determine the flood loads to use in the determination of F_a   Table 2.11

- Global foundation loads

    For the analysis of the entire (global) foundation, it is reasonable to assume that one of the 20 columns will be subjected to the impact load, *F*, and the remaining columns will be subjected to the hydrodynamic load, F_{dyn}. Therefore, the total horizontal flood load on the foundation is equal to the following where the normal impact load is greater than the special impact load:

$$F_a = F + (n-1)F_{dyn} = 5,099 + (19 \times 860) = 21,439 \text{ lb}$$

In S.I.:

$$F_a = 23.0 + (19 \times 3.8) = 95.2 \text{ kN}$$

The overturning moment about *GS* due to the flood loads is equal to the following:

$$F_a = Fd_s + (n-1)F_{dyn}(d_s/2) = (5,099 \times 3.0) + (19 \times 860 \times 3.0/2) = 39,807 \text{ ft-lb}$$

In S.I.:

$$F_a = (23.0 \times 0.91) + (19 \times 3.8 \times 0.91/2) = 53.8 \text{ kN-m}$$

- Individual column loads

    The flood loads on an individual column in the front row consists of *F* and the greater of F_{dyn} and F_D (see Fig. 2.21). The load effects due to these loads are combined with the other applicable loads using the appropriate load combinations in Sec. 2.16.2 of this publication.

The design and construction of the mat foundation must satisfy the requirements of ASCE/SEI 24 Section 4.5. The top of the mat must be located below the *GS* and must extend to a depth sufficient to prevent flotation, collapse, or permanent lateral movement under the design load combinations (ASCE/SEI 24 Section 1.5.3).

2.17.4   Example 2.4—Piles Supporting a Utility Building Located in a VE Zone

Determine the flood loads, F_a, on the piles supporting the utility building depicted in Fig. 2.22. The location of the building is noted on the FIRM in Fig. 2.23. Assume the following:

- Source of the floodwaters is from coastal waters, which is salt water.
- The ground slopes up gently from the shoreline and there are nominal obstructions between the shoreline and the site.

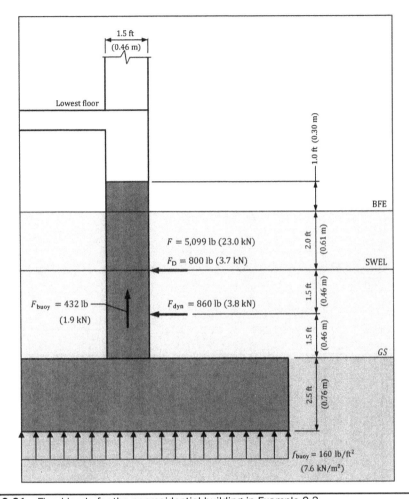

FIGURE 2.21 Flood loads for the nonresidential building in Example 2.3.

- The utility building must remain operational before and after a design flood event and is oriented perpendicular to the shoreline.
- From the FIRM, the flood hazard Zone is VE and BFE = 7.0 ft NAVD (2.13 m NAVD).
- From the FIS, SWEL = 4.0 ft NAVD (1.22 m NAVD).
- GS = 0.0 ft NAVD (0.0 m NAVD).
- The local jurisdiction requires a 1.0 ft (0.30 m) freeboard.
- Floodwater debris hazards exist and are characterized as normal and special.

Solution

Step 1—Acquire the flood hazard Zone

From the design data, the building is located in a VE Zone.

Step 2—Acquire the base flood elevation, BFE

From the FIRM, BFE = 7.0 ft NAVD (2.13 m NAVD).

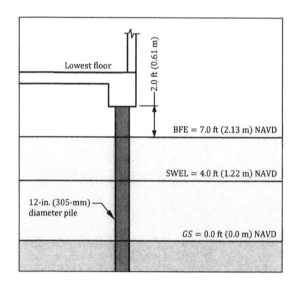

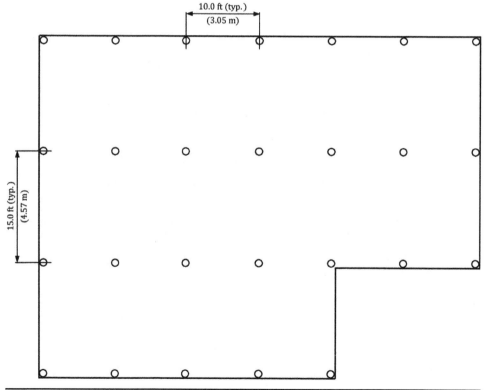

Figure 2.22 Utility building in Example 2.4.

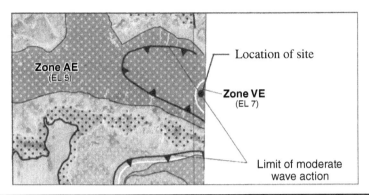

FIGURE 2.23 Location of the utility building in Example 2.4.

Step 3—Determine the design flood elevation, DFE

Because the local jurisdiction has adopted a more severe flood than that obtained from the FIRM, DFE = BFE + freeboard = 7.0 + 1.0 = 8.0 ft NAVD (2.44 m NAVD).

Step 4—Determine the minimum elevation of the lowest floor   Fig. 2.3

Because the utility building must remain operational during and after a design flood event, the flood design class is 3 (see Table 1-1 in ASCE/SEI 24). Therefore, for a VE Zone, the minimum elevation of the bottom of the lowest supporting horizontal structural member is equal to the higher of the following:

- BFE + 2.0 = 7.0 + 2.0 = 9.0 ft NAVD (2.74 m NAVD)
- DFE = 8.0 ft NAVD (2.44 m NAVD)

Step 5—Determine the stillwater depth, d_s

$$d_s = SWEL - GS = 4.0 - 0.0 = 4.0 \text{ ft (1.22 m)}$$

Step 6—Determine the breaking wave height, H_b, where applicable   Fig. 2.5

For a gently sloping shoreline with nominal obstructions:

$$H_b = 0.78d_s = 0.78 \times 4.0 = 3.1 \text{ ft (0.95 m)} \qquad \text{Eq. (2.1)}$$

Step 7—Determine the design flood velocity, V   Table 2.3

Because the building is located in a VE Zone, the upper bound average water velocity is appropriate:

$$V = (gd_s)^{0.5} = (32.2 \times 4.0)^{0.5} = 11.3 \text{ ft/s}$$

In S.I.:

$$V = (gd_s)^{0.5} = (9.81 \times 1.22)^{0.5} = 3.46 \text{ m/s}$$

Step 8—Determine the hydrostatic loads, F_{sta} and F_{buoy}, where applicable   Fig. 2.8

Lateral hydrostatic loads on the piles are included in the hydrodynamic load, F_{dyn}.

There is no buoyancy load on the lowest floor slab.

The buoyancy load on a pile is equal to the following:

$$F_{buoy} = \gamma_w d_s A_p = 64.0 \times 4.0 \times \pi \times (12.0/12)^2/4 = 201 \text{ lb}$$

In S.I.:

$$F_{buoy} = \gamma_w d_s A_c = 10.05 \times 1.22 \times \pi \times (305/1,000)^2/4 = 0.90 \text{ kN}$$

Step 9—Determine the hydrodynamic load, F_{dyn}, where applicable   Fig. 2.9

Hydrodynamic loads occur on the piles and are determined by the following equation:

$$F_{dyn} = \frac{1}{2} a\rho V^2 A \qquad \text{Eq. (2.5)}$$

From Fig. 2.9, use $a = 1.25$ for round piles.

Therefore, F_{dyn} on one of the piles is equal to the following:

$$F_{dyn} = \frac{a\rho V^2 D d_s}{2} = \frac{1.25 \times 1.99 \times 11.3^2 \times (12.0/12) \times 4.0}{2} = 635 \text{ lb}$$

This load acts at $d_s/2 = 4.0/2 = 2.0$ ft above GS.

In S.I.:

$$F_{dyn} = \frac{a\rho V^2 D d_s}{2} = \frac{1.25 \times 1.026 \times 3.46^2 \times (305/1,000) \times 1.22}{2} = 2.9 \text{ kN}$$

This load acts at $d_s/2 = 1.22/2 = 0.61$ m above GS.

Step 10—Determine breaking wave loads, where applicable   Fig. 2.11

Breaking wave loads, F_D, occur on the vertical piles and are determined by ASCE/SEI Equation (5.4-4):

$$F_D = 0.5\gamma_w C_D D H_b^2$$

$C_D = 1.75$ for round piles.   ASCE/SEI 5.4.4.1

$D = 1.0$ ft   ASCE/SEI 5.4.4.1

Therefore, F_D on one of the piles is equal to the following:

$$F_D = 0.5 \times 64.0 \times 1.75 \times 1.0 \times 3.1^2 = 538 \text{ lb}$$

This load acts at the SWEL, which is 4.0 ft above GS.

In S.I.:

$$D = 0.3 \text{ m}$$

$$F_D = 0.5 \times 10.05 \times 1.75 \times (305/1,000) \times 0.95^2 = 2.4 \text{ kN}$$

This load acts at the SWEL, which is 1.22 m above GS.

Step 11—Determine normal and special impact loads, F, where applicable

Normal impact loads are determined using the flowchart in Fig. 2.12.

- Determine the debris impact weight, W

  For coastal areas where piers and large pilings are present, use $W = 2,000$ lb (9.2 kN) [see the table in Fig. 2.12].

- Determine the debris velocity, V_b

  It is reasonable to assume $V_b = V = 11.3$ ft/s (3.46 m/s).

- Determine the importance coefficient, C_I

  For a Risk Category III building:

$$C_I = 1.2 \qquad\qquad\qquad \text{ASCE/SEI Table C5.4-1}$$

- Orientation coefficient, $C_O = 0.8$.   ASCE/SEI C5.4.5
- Determine the depth coefficient, C_D

  For a VE Zone:

$$C_D = 1.0$$

- Determine the blockage coefficient, C_B

  Assuming there is no upstream screening and the flow path is wider than 30 ft (9.14 m):

$$C_B = 1.0 \qquad\qquad\qquad \text{ASCE/SEI Table C5.4-3}$$

- Determine the natural period, T_n, of the utility building.

  From a dynamic analysis, $T_n = 0.30$ s.

- Determine the ratio of the impact duration to the natural period of structure, $\Delta t / T_n$.

  Given the impact duration $\Delta t = 0.03$ s:

$$\Delta t / T_n = 0.03/0.30 = 0.10$$

- Determine the maximum response ratio for impulsive loads, R_{max}

  For $\Delta t / T_n = 0.10$:

$$R_{max} = 0.4 \qquad\qquad\qquad \text{ASCE/SEI Table C5.4-4}$$

- Determine the normal impact force, F ASCE/SEI Equation (C5.4-3)

$$F = \frac{\pi W V_b C_I C_O C_D C_B R_{max}}{2g\Delta t}$$

$$= \frac{\pi \times 2,000 \times 11.3 \times 1.2 \times 0.8 \times 1.0 \times 1.0 \times 0.4}{2 \times 32.2 \times 0.03} = 14,112 \text{ lb}$$

In S.I.:

$$F = \frac{\pi WV_b C_1 C_O C_D C_B R_{max}}{2g\Delta t}$$

$$= \frac{\pi \times 9.0 \times 3.46 \times 1.2 \times 0.8 \times 1.0 \times 1.0 \times 0.4}{2 \times 9.81 \times 0.03} = 63.8 \text{ kN}$$

This load acts at the SWEL, which is 4.0 ft (1.22 m) above GS. Special impact loads are determined by Eq. (2.8):

$$F = \frac{C_D \rho A V^2}{2} = \frac{1.0 \times 1.99 \times [(12.0/12) \times 4.0] \times 11.3^2}{2} = 508 \text{ lb}$$

In S.I.:

$$F = \frac{C_D \rho A V^2}{2} = \frac{1.0 \times 1.026 \times (0.305 \times 1.22) \times 3.46^2}{2} = 2.3 \text{ kN}$$

This load acts at the SWEL, which is 4.0 ft (1.22 m) above GS.

Step 12—Determine the flood loads to use in the determination of F_a　　　Table 2.11

- Global foundation loads

For the analysis of the entire (global) foundation, it is reasonable to assume that one of the 26 piles will be subjected to the impact load, F, and the remaining piles will be subjected to the hydrodynamic load, F_{dyn}. Therefore,

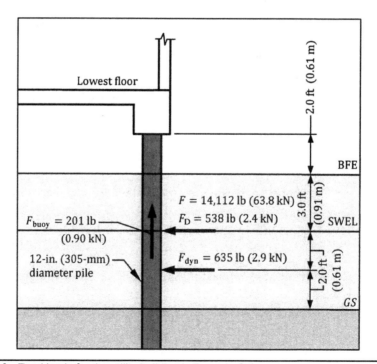

FIGURE 2.24 Flood loads for the utility building in Example 2.4.

the total horizontal flood load on the foundation is equal to the following where the normal impact load is greater than the special impact load:

$$F_a = F + (n-1)F_{dyn} = 14,112 + (25 \times 635) = 29,987 \text{ lb}$$

In S.I.:

$$F_a = 63.8 + (25 \times 2.9) = 136.3 \text{ kN}$$

The overturning moment about GS due to the flood loads is equal to the following:

$$F_a = Fd_s + (n-1)F_{dyn}(d_s/2)$$

$$= (14,112 \times 4.0) + (25 \times 635 \times 4.0/2) = 88,198 \text{ ft-lb}$$

In S.I.:

$$F_a = (63.8 \times 1.22) + (25 \times 2.9 \times 1.22/2) = 122.1 \text{ kN-m}$$

- Individual pile loads

  The flood loads on an individual pile in the front row consists of F and the greater of F_{dyn} and F_D (see Fig. 2.24). The load effects due to these loads are combined with the other applicable loads using the appropriate load combinations in Sec. 2.16.2 of this publication.

CHAPTER 3
Tsunami Loads

3.1 Overview

This chapter contains methods to calculate design tsunami loads and effects, F_{TSU}. Certain buildings and structures located in tsunami-prone regions (TPRs) must be designed and constructed to resist the effects of tsunamis in accordance with Chapter 6 of ASCE/SEI 7 (IBC 1615.1).

Tsunami requirements are applicable to the states of Alaska, California, Hawaii, Oregon, and Washington. Their coastlines have designated TPRs where quantifiable probabilistic hazards resulting from tsunamigenic earthquakes with subduction faulting make them vulnerable to being flooded or inundated by a tsunami.

Tsunamis have the potential of generating very large hydrodynamic loads and it has been documented that one- and two-family dwellings and low-rise buildings of light-frame construction typically do not survive a design tsunami event. Therefore, it is not practical to provide tsunami design requirements for these types of structures (ASCE/SEI C6.1.1).

3.2 Notation

A_{beam} = vertical projected area of an individual beam element, ft² (m²)

A_{col} = vertical projected area of an individual column element, ft² (m²)

A_d = vertical projected area of obstructing debris accumulated on a structure, ft² (m²)

A_{wall} = vertical projected area of an individual wall element, ft² (m²)

B = overall building width, ft (m)

b = width of component subject to force, ft (m)

C_{cx} = closure coefficient

C_d = drag coefficient based on quasi-steady loads

C_o = orientation coefficient for debris

D = dead load

D_s = scour depth, ft (m)

DT = displacement tonnage

DWT = deadweight tonnage of vessel

E_g = hydraulic head in the Energy Grade Line Analysis, ft (m)

$E_{g,i}$ = hydraulic head at point i in the Energy Grade Line Analysis, ft (m)

73

E_{mh} = horizontal seismic load effect including the system overstrength factor, defined in ASCE/SEI 12.4.3.1

F_{dx} = drag force on the building or element at each level, lb (kN)

F_h = unbalanced hydrostatic lateral force, lb (kN)

F_i = debris impact force, lb (kN)

F_{ni} = nominal maximum instantaneous debris impact force, lb (kN)

F_{pw} = hydrodynamic force on a perforated wall, lb (kN)

F_r = Froude number = $u/\sqrt{gh}$

F_{TSU} = tsunami load or effect

F_v = buoyancy force, lb (kN)

F_w = lateral hydrodynamic force on a wall or pier, lb (kN)

$F_{w\theta}$ = lateral hydrodynamic force on a wall oriented at an angle θ to the flow, lb (kN)

F_x = drag force on an element or component, lb (kN)

f_{uw} = equivalent uniform lateral force per unit width, lb/ft (kN/m)

g = acceleration due to gravity = 32.2 ft/s² (9.81 m/s²)

H_T = offshore tsunami amplitude, ft (m) (see ASCE/SEI Figure 6.7-1)

H_{TSU} = load caused by tsunami-induced lateral earth pressure under submerged conditions, lb (kN)

h = tsunami inundation depth above the grade plane at the structure, ft (m)

h_e = inundation height of an individual element, ft (m)

h_i = inundation depth at point i, ft (m)

h_{max} = maximum inundation depth at the building or other structure, ft (m)

h_o = offshore water depth, ft (m)

h_r = residual water height within a building, ft (m)

h_s = height of structural floor slab above the grade plane at the structure, ft (m)

h_{sx} = story height of story x, ft (m)

I_{tsu} = importance factor for tsunami forces to account for additional uncertainty in estimated parameters

k_s = fluid density factor to account for suspended soil and other smaller flow-embedded objects that are not considered in ASCE/SEI 6.11 = 1.1

L = live load

ℓ_w = length of a structural wall, ft (m)

LWT = lightship weight of vessel

$m_{contents}$ = mass of contents in a shipping container

NAVD = North American Vertical Datum of 1988

n = Manning's coefficient

P_u = uplift pressure on a slab or building horizontal element, lb/ft² (kN/m²)

P_{ur} = reduced uplift pressure for a slab with an opening, lb/ft² (kN/m²)

p_r = residual water surcharge pressure, lb/ft^2 (kN/m^2)

p_s = hydrostatic surcharge pressure due to tsunami inundation, lb/ft^2 (kN/m^2)

p_{uw} = equivalent maximum uniform pressure accounting for unbalanced lateral hydrostatic and hydrodynamic loads, lb/ft^2 (kN/m^2)

R = mapped tsunami runup elevation, ft (m)

R_{max} = dynamic response ratio

S = snow load

s = friction slope of the energy grade line

T_{TSU} = predominant wave period, or the time from the start of the first pulse to the end of the second pulse, s

t = time, s

t_d = duration of debris impact, s

TDZ = tsunami design zone

u = tsunami flow velocity, ft/s (m/s)

u_{max} = maximum tsunami flow velocity at the structure, ft/s (m/s)

u_v = vertical component of the tsunami flow velocity, ft/s (m/s)

V = seismic base shear determined in accordance with ASCE/SEI Chapter 12

V_w = displaced water volume, ft^3 (m^3)

W_s = weight of the structure, lb (kN)

x_i = horizontal distance measured from the runup point, ft (m)

x_R = mapped inundation limit distance inland from the NAVD shoreline, ft (m)

z = ground elevation above the NAVD datum, ft (m)

α = Froude number coefficient in the Energy Grade Line Analysis

γ_s = minimum fluid weight density for design hydrostatic loads, lb/ft^3 (kN/m^3)

γ_{sw} = effective weight density of seawater
     = 64.0 lb/ft^3 (10.05 kN/m^3)

Δx_i = incremental horizontal distance used in the Energy Grade Line Analysis, ft (m)

ξ_{100} = surf similarity parameter using 328 ft (100 m) nearshore wave characteristics

θ = angle between the longitudinal axis of a wall and the flow direction

ρ_s = minimum fluid mass density for design hydrodynamic loads, $slugs/ft^3$ (kg/m^3)

ρ_{sw} = effective mass density of seawater
     = 1.99 $slugs/ft^3$ (1,025 kg/m^3)

ϕ = structural resistance factor

φ = average slope of grade at the structure

φ_i = average slope of grade at point i

Φ = mean slope angle of the nearshore profile

Ω = angular frequency of the waveform

Ω_o = overstrength factor for the lateral-force-resisting system given in ASCE/SEI Table 12.2-1

3.3 Terminology

Terminology used throughout this chapter related to tsunami loads is given in Table 3.1.

Bathymetric profile	A cross section showing ocean depth plotted as a function of horizontal distance from a reference point (such as a coastline).
Critical facility	• Buildings and structures providing services designated by federal, state, local, or tribal governments to be essential for the implementation of the response and management plan or for the continued service functioning of a community (such as, facilities for power, fuel, water, communications, public health, major transportation infrastructure, and essential government operations). • Critical facilities comprise all public and private facilities deemed by a community to be essential for the delivery of vital services, protection of special populations, and the provision of other services of importance for that community.
Deadweight tonnage (DWT)	• A vessels displacement tonnage (DT) minus its lightship weight (LWT). • DWT is a classification used for the carrying capacity of a vessel, which is the sum of the weights of cargo, fuel, fresh water, ballast water, provisions, passengers, and crew. The weight of the vessel itself is not included in the DWT.
Design tsunami parameters	Tsunami parameters used for design, consisting of the inundation depths and flow velocities at the stages of inflow and outflow most critical to the structure and momentum flux.
Displacement tonnage	Total weight of a fully loaded vessel.
Grade plane	• A horizontal reference plane at the site representing the average elevation of finished ground level adjoining the structure at all exterior walls. • Where the finished ground level slopes away from the exterior walls and the property line is less than or equal to 6 ft (1.8 m) from the structure, the grade plane is established by the lowest points within the area between the structure and the property line. • Where the finished ground slopes away from the exterior walls and the property line is more than 6 ft (1.8 m) from the structure, the grade plane is established by the lowest points within the area between the structure and points 6 ft (1.8 m) from the structure.
Inundation depth	The depth of the design tsunami water level, including relative sea level change, with respect to the grade plane at the structure.
Inundation limit	• The maximum horizontal inland extent of flooding for the maximum considered tsunami (MCT), where the inundation depth above grade becomes zero. • The horizontal distance that is flooded relative to the shoreline defined where the North American Vertical Datum (NAVD) of 1988 elevation is zero.
Lightship weight	Weight of a vessel without cargo, cargo, fuel, fresh water, ballast water, provisions, passengers, and crew.
Maximum considered tsunami	A probabilistic tsunami having a 2 percent probability of being exceeded in a 50-year period (that is, a 2,475-year mean recurrence interval).

TABLE 3.1 Terminology Related to Tsunami Loads

Nearshore profile	Cross-sectional bathymetric profile from the shoreline to a water depth of 328 ft (100 m).
Nearshore tsunami amplitude	MCT amplitude immediately off the coastline at 33 ft (10 m) of water depth.
Offshore tsunami Amplitude	MCT amplitude relative to the reference sea level, measured where the undisturbed water depth is 328 ft (100 m).
Offshore tsunami height	Waveform vertical dimension of the MCT from consecutive trough to crest, measured where the undisturbed water depth is 328 ft (100 m), after removing the tidal variation.
Reference sea level	The sea level datum used in site-specific inundation modeling, which is typically taken as the mean high water level (MHWL).
Runup elevation	Ground elevation at the maximum tsunami inundation limit, including relative sea level change, with respect to the NAVD of 1988 reference datum.
Topographic transect	Profile of vertical elevation data versus horizontal distance along a cross section of the terrain, in which the orientation of the cross section is perpendicular or at some specified orientation angle to the shoreline.
Tsunami	A series of waves with variable long periods, typically resulting from earthquake-induced uplift or subsidence of the seafloor.
Tsunami amplitude	The absolute value of the difference between a particular peak or trough of the tsunami and the undisturbed sea level at the time.
Tsunami breakaway wall	A nonstructural wall that is designed to fail when subjected to tsunami loads.
Tsunami bore	A steep and turbulent broken wave front generated on the front edge of a long-period tsunami waveform.
Tsunami bore height	The height of a broken tsunami surge above the water level in front of the bore or grade elevation if the bore arrives on nominally dry land.
Tsunami design zone (TDZ)	An area identified on the Tsunami Design Zone Map (TDZM) between the shoreline and the inundation limit, within which buildings and structures are analyzed and designed for inundation by the MCT.
Tsunami design zone map	• The map in ASCE/SEI Figure 6.1-1 designating the potential horizontal inundation limit of the MCT. • A state or local jurisdiction's probabilistic map produced in accordance with the requirements of ASCE/SEI 6.7.
Tsunami-prone region	The coastal region in the United States addressed by ASCE/SEI Chapter 6 with quantified probability in the recognized literature of tsunami inundation hazard with a runup greater than 3 ft (0.91 m) caused by tsunamigenic earthquakes in accordance with the Probabilistic Tsunami Hazard Analysis (PTHA) method given in ASCE/SEI Chapter 6.
Tsunami risk category (TRC)	The risk category in ASCE/SEI 1.5, as modified for specific use related to ASCE/SEI Chapter 6 in accordance with ASCE/SEI 6.4.
Tsunami vertical evacuation refuge structure (TVERS)	A structure designated and designed to serve as a point of refuge to which a portion of the community's population can evacuate above a tsunami when high ground is not available.

TABLE 3.1 Terminology Related to Tsunami Loads (*Continued*)

3.4   Procedure to Determine Tsunami Loads, F_{TSU}

A step-by-step procedure to determine tsunami loads, F_{TSU}, is given in Fig. 3.1. It is assumed the building or structure is located in a TDZ in accordance with the TDZM of the local jurisdiction. The section, table, and figure numbers of this publication referenced in Fig. 3.1 contain additional information to calculate F_{TSU}.

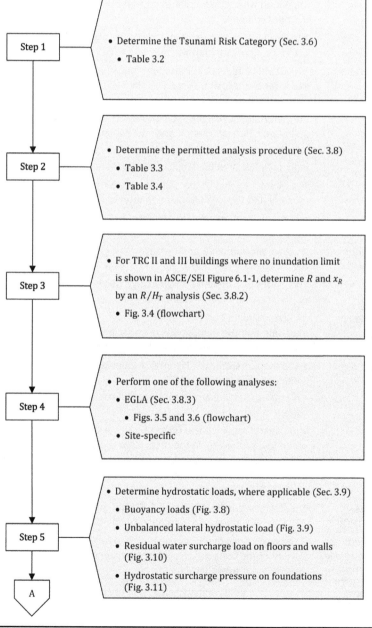

Figure 3.1   Procedure to determine tsunami loads, F_{TSU}.

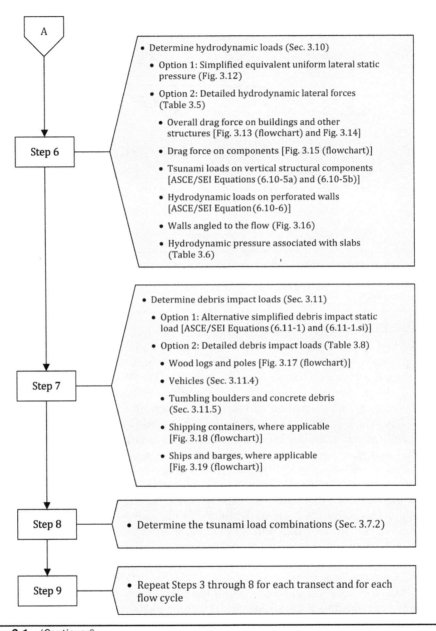

A

Step 6

- Determine hydrodynamic loads (Sec. 3.10)
  - Option 1: Simplified equivalent uniform lateral static pressure (Fig. 3.12)
  - Option 2: Detailed hydrodynamic lateral forces (Table 3.5)
    - Overall drag force on buildings and other structures [Fig. 3.13 (flowchart) and Fig. 3.14]
    - Drag force on components [Fig. 3.15 (flowchart)]
    - Tsunami loads on vertical structural components [ASCE/SEI Equations (6.10-5a) and (6.10-5b)]
    - Hydrodynamic loads on perforated walls [ASCE/SEI Equation (6.10-6)]
    - Walls angled to the flow (Fig. 3.16)
    - Hydrodynamic pressure associated with slabs (Table 3.6)

Step 7

- Determine debris impact loads (Sec. 3.11)
  - Option 1: Alternative simplified debris impact static load [ASCE/SEI Equations (6.11-1) and (6.11-1.si)]
  - Option 2: Detailed debris impact loads (Table 3.8)
    - Wood logs and poles [Fig. 3.17 (flowchart)]
    - Vehicles (Sec. 3.11.4)
    - Tumbling boulders and concrete debris (Sec. 3.11.5)
    - Shipping containers, where applicable [Fig. 3.18 (flowchart)]
    - Ships and barges, where applicable [Fig. 3.19 (flowchart)]

Step 8

- Determine the tsunami load combinations (Sec. 3.7.2)

Step 9

- Repeat Steps 3 through 8 for each transect and for each flow cycle

FIGURE 3.1   *(Continued)*

A summary of the tsunami parameters used in the determination of tsunami loads is given in Fig. 3.2.

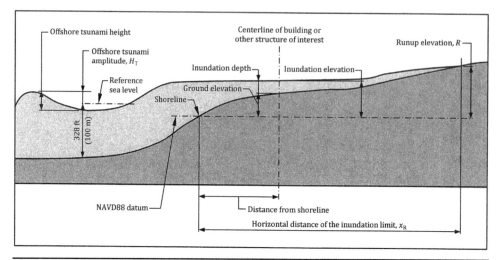

FIGURE 3.2   Tsunami parameters for a topographic flow transect.

3.5   Tsunami Design Zone Map

The first step is to determine if the site is located in a TPRs. The TDZM in ASCE/SEI Figure 6.1-1 designates the potential horizontal inundation limits of the MCT for Alaska, California, Hawaii, Oregon, and Washington. The TDZ is an area identified on the TDZM between the shoreline and the inundation limit within which buildings and structures must be analyzed and designed for inundation by the MCT.

The ASCE Tsunami Design Geodatabase (abbreviated hereafter as ASCE Geodatabase) contains maps depicting the extent of inundation (that is, the TDZ) due to the MCT and is an integral part of the tsunami design provisions of ASCE/SEI Chapter 6 (https://asce7tsunami.online/). A site can be located by address, by latitude or longitude, or by clicking on the map. In addition to the TDZ (which is shaded light blue on the maps), offshore tsunami amplitudes, runup elevation points, and inundation depth points are given (which are designated by blue squares, red triangles, and red circles, respectively). Earthquake-induced regional ground subsidence contours are indicated as well. The distance from the shoreline to the inundation limit can be obtained by utilizing the "Transect" or "Measure" features of the program.

A state or local jurisdiction may have adopted a TDZ and other data different from those in the ASCE Geodatabase, so it is good practice to check with the local authorities prior to performing the analysis.

In addition to the ASCE Geodatabase, ASCE 7-16 Tsunami Design Zone Maps for Selected Locations (http://ascelibrary.org/doi/book/10.1061/9780784480748) contains tsunami design zone maps in high-resolution PDF format for the 62 locations indicated by circles in ASCE/SEI Figures. C6.1-1(a) to (i). These downloadable maps can be used to quickly determine whether a structure falls within the TDZ, which is indicated on the maps by red shading. For the purpose of identifying the TDZ, these maps are considered equivalent to the results obtained from the ASCE Geodatabase.

Sea level rise is not incorporated into the TDZM. The runup elevation and the inundation depth for a given MCT must be increased by at least the future projected relative sea level rise during the design life of a structure. A project life cycle of not less than 50 years should be considered (ASCE/SEI 6.5.3). Relative sea level change at a site may result from ground subsidence, long-term erosion, thermal expansion caused by warming of the ocean, or increased melting of land-based ice. Sea level changes can be obtained from the NOAA Center for Operational Oceanographic Products and Services (https://tidesandcurrents .noaa.gov/sltrends/). Once the potential increase in relative sea level has been determined, it must be added to the inundation depth results of the Energy Grade Line Analysis (EGLA).

3.6 Tsunami Risk Categories and General Requirements

The following buildings and other structures located within a TDZ must be analyzed and designed for the MCT (ASCE/SEI 6.1.1):

- TRC IV buildings and structures.
- TRC III buildings and structures with an inundation depth greater than 3 ft (0.91 m) at any location within the intended footprint of the structure.
- Where required by a state or locally adopted building code statue to include design for tsunamis effects, TRC II buildings with a mean height above the grade plane greater than the height designated in the statue and with an inundation depth greater than 3 ft (0.91 m) at any location within the intended footprint of the building.

Two exceptions to these requirements are given in ASCE/SEI 6.1.1:

- Tsunami loads and effects determined in accordance with ASCE/SEI Chapter 6 need not be applied to TRC II single-story buildings of any height without mezzanines or any occupiable roof level and not having any critical equipment or systems.
- For coastal regions subject to tsunami inundation and not covered in ASCE/SEI Figure 6.1-1, the TDZ, inundation limits, and runup elevations must be determined by either (1) the site-specific procedures of ASCE/SEI 6.7 or (2) the procedures in ASCE/SEI 6.5.1.1 using ASCE/SEI Figure 6.7-1 for TRC II or III structures.

TRCs are the same as the risk categories defined in ASCE/SEI 1.5 with the following modifications (ASCE/SEI 6.4; see Table 3.2):

1. Federal, state, local, or tribal governments are permitted to include critical facilities in TRC III, such as power-generating stations, water-treatment facilities for potable water, waste-water treatment facilities, and other public utility facilities not included in Risk Category IV. Additional information on this modification can be found in ASCE/SEI C6.4.

2. The following structures need not be included in TRC IV, and state, local, or tribal governments are permitted to designate them to TRC II or III:

   a. Fire stations, ambulance facilities, and emergency vehicle garages;

   b. Earthquake or hurricane shelters;

   c. Emergency aircraft hangars; and

   d. Police stations without holding cells and not uniquely required for post-disaster emergency response as a critical facility.

3. TVERS must be included in TRC IV.

Risk Category	Use or Occupancy
I	Buildings and other structures that represent a low risk to humans.
II	All buildings and other structures except those listed in Risk Categories I, III, and IV.
III	• Buildings and other structures, the failure of which could pose a substantial risk to human life. • Buildings and other structures with potential to cause a substantial economic impact and/or mass disruption of day-to-day civilian life in the event of failure.
IV	• Buildings and other structures designated as essential facilities. • Buildings and other structures, the failure of which could pose a substantial hazard to the community.

TABLE 3.2 Summary of Risk Categories of Buildings and Other Structures (ASCE/SEI Table 1.5-1)

The essential facilities listed in Item II above do not need to be included in TRC IV because these facilities are expected to be evacuated prior to the arrival of a tsunami. Additional information on these modifications can be found in ASCE/SEI C6.4.

A TVERS is a structure designated and designed to serve as a point of refuge where people can evacuate above a tsunami (ASCE/SEI 6.14). More information on such structures is given in ASCE/SEI C6.14.

3.7 Structural Performance Evaluation

3.7.1 Load Cases

The three load cases in ASCE/SEI 6.8.3.1 must be evaluated for any building or structure (see Fig. 3.3). The hydrodynamic force, F_{dx}, is calculated by ASCE/SEI Equation (6.10-2) based on the conditions in each load case. These forces and other applicable forces are combined in accordance with the load combinations in ASCE/SEI 6.8.3.3 and are applied to the lateral force–resisting system of the building.

In load case 1, the structure is subjected to the maximum buoyant force, F_v, and the hydrodynamic force, F_{dx}, where it is assumed that the exterior walls have not failed; this means the interior of the building is still dry, resulting in uplift forces due to buoyancy. The purpose of this load case is to check the stability of the structure and its foundation for uplift. It need not be applied to open structures nor to structures where the soil properties or foundation and structural design prevent detrimental hydrostatic pressurization on the underside of the foundation and the lowest structural slab.

Maximum flow velocity and two-thirds the maximum inundation depth are used in load case 2. This load case results in the maximum hydrodynamic forces on the lateral force–resisting system of the structure.

In load case 3, the maximum inundation depth and one-third the maximum flow velocity are used to determine the hydrodynamic forces.

Unless a site-specific tsunami analysis is performed in accordance with ASCE/SEI 6.7, the inundation depths and velocities defined for the load cases must be determined using the time-history graphs in ASCE/SEI Figure 6.8-1 (numerical values of normalized inundation depths and flow velocities from the graphs are given in ASCE/SEI

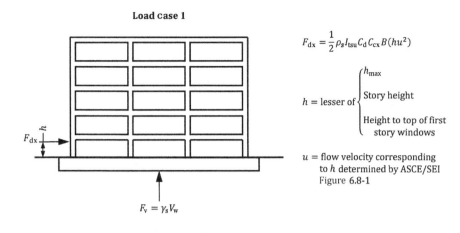

Load case 1

$$F_{dx} = \frac{1}{2}\rho_s I_{tsu} C_d C_{cx} B(hu^2)$$

$$h = \text{lesser of} \begin{cases} h_{max} \\ \text{Story height} \\ \text{Height to top of first story windows} \end{cases}$$

u = flow velocity corresponding to h determined by ASCE/SEI Figure 6.8-1

$$F_v = \gamma_s V_w$$

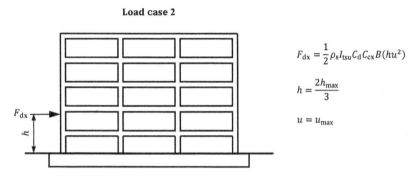

Load case 2

$$F_{dx} = \frac{1}{2}\rho_s I_{tsu} C_d C_{cx} B(hu^2)$$

$$h = \frac{2h_{max}}{3}$$

$$u = u_{max}$$

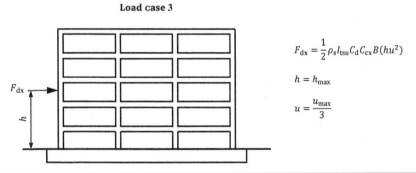

Load case 3

$$F_{dx} = \frac{1}{2}\rho_s I_{tsu} C_d C_{cx} B(hu^2)$$

$$h = h_{max}$$

$$u = \frac{u_{max}}{3}$$

Figure 3.3 Tsunami load cases.

Table C6.8-2). In the top graph, normalized inundation depths, h/h_{max}, are plotted versus normalized time, t/T_{TSU}, where T_{TSU} is the predominant wave period obtained from the ASCE Geodatabase. Load cases 2 and 3 are identified in the graph for both incoming and receding directions of flow. In the bottom graph, normalized flow velocities, u/u_{max}, are plotted versus t/T_{TSU}. The intent of the graphs is to capture all possible combinations of inundation depth and velocity flow during the time sequence of a tsunami inundation.

3.7.2   Load Combinations

Principal tsunami forces and effects, F_{TSU}, are to be combined with other applicable forces and effects using the following load combinations [see ASCE/SEI Equations (6.8-1a) and (6.8-1b)]:

$$0.9D + F_{TSU} + H_{TSU} \tag{3.1}$$

$$1.2D + F_{TSU} + 0.5L + 0.2S + H_{TSU} \tag{3.2}$$

In these equations, H_{TSU} is the load caused by the tsunami-induced lateral foundation pressures developed under submerged conditions. The load factor on H_{TSU} must be taken as 0.9 where the net effect of H_{TSU} counteracts the principal load effect. The resistance factors for foundation stability analysis, ϕ, are given in ASCE/SEI 6.12.1.

These load combinations are consistent with those given in ASCE/SEI 2.5 for extraordinary events.

3.7.3   Lateral Force–Resisting System Acceptance Criteria

For buildings and other structures assigned to Seismic Design Category (SDC) D, E, or F in accordance with ASCE/SEI 11.6 where the capacity of the structural system is required to be evaluated at the life safety structural performance level (that is, the post-tsunami damage state in which a structure has damaged components but retains a margin against the onset of partial or total collapse), it is permitted to evaluate the system using $0.75E_{mh} = 0.75\Omega_o V$ (ASCE/SEI 6.8.3.4). The term E_{mh} is the horizontal seismic load effect including the system overstrength factor, Ω_o, and V is the horizontal seismic base shear determined in accordance with ASCE/SEI Chapter 12. Structures typically resist the effects from the design earthquake prior to the effects from the design tsunami, and the overall structural capacity is considered acceptable without additional analyses provided the maximum tsunami force acting on the structure is less than or equal to $0.75\Omega_o V$. In other words, the lateral force–resisting system does not have sufficient capacity to resist a maximum tsunami load greater than $0.75\Omega_o V$, and the structural members must be designed and detailed accordingly.

The lateral force–resisting system for buildings and other structures assigned to SDC D and above must be explicitly analyzed and evaluated for acceptance where the immediate occupancy structural performance criteria must be satisfied.

3.7.4   Structural Component Acceptance Criteria

Structural components must be designed for the load effects that result from the overall tsunami forces on the structural system combined with any resultant actions caused by local tsunami pressures for a given flow direction. In particular, the design of structural components must comply with the acceptability criteria of ASCE/SEI 6.8.3.5.1, the performance-based criteria of ASCE/SEI 6.8.3.5.2, or the progressive collapse avoidance criteria of ASCE/SEI 6.8.3.5.3.

According to the acceptability criteria in ASCE/SEI 6.8.3.5.1, components must be designed for combined load effects using a linearly elastic, static analysis and the load combinations in ASCE/SEI 6.8.3.3. The material resistance factors, ϕ, that must be used are those prescribed in the material-specific standards, like ACI 318 for reinforced-concrete members.

Information on the proper application of the criteria of ASCE/SEI 6.8.3.5.2 and ASCE/SEI 6.8.3.5.3 is given in ASCE/SEI C6.8.3.5.2 and ASCE/SEI C6.8.3.5.3, respectively.

3.8   Analysis Procedures to Determine Tsunami Inundation Depth and Flow Velocity

3.8.1   Permitted Analysis Procedures

The design inundation depth and flow velocity at a site are needed to calculate design tsunami loads. The applicable method to determine these parameters depends on the TRC assigned to the building or other structure and whether the inundation limit and runup elevation of the MCT can be determined by ASCE/SEI Figure 6.1-1 or not.

Tsunami Risk Category II and II Buildings and Other Structures

Case 1: Inundation depth and runup elevation are given in ASCE/SEI Figure 6.1-1
Where information on inundation depth and runup elevation is given in ASCE/SEI Figure 6.1-1 for a particular site, maximum inundation depths and flow velocities must be determined by the EGLA in ASCE/SEI 6.6. The site-specific Probabilistic Tsunami Hazard Analysis (PTHA) in ASCE/SEI 6.7 is permitted as an alternate to the EGLA. The flow velocities determined by the PTHA are subject to the limitations in ASCE/SEI 6.7.6.8.

Case 2: No mapped inundation limit is given in ASCE/SEI Figure 6.1-1 and the offshore tsunami amplitude is given in ASCE/SEI Figure 6.7-1
Where no mapped inundation limit is given in ASCE/SEI Figure 6.1-1 and the off-shore tsunami amplitude (which is used to determine the runup elevation) is given in ASCE/SEI Figure 6.7-1, the R/H_T analysis method in ASCE/SEI 6.5.1.1 is permitted to be used for TRC II and III buildings and other structures.

Tsunami Risk Category IV Buildings and Other Structures

The EGLA and the site-specific PTHA must both be performed for TRC IV buildings and other structures. However, for structures other than TVERS, a site-specific analysis need not be performed where the inundation depth determined by the EGLA method is less than 12 ft (3.7 m) at any point within the footprint of the structure.

Site-specific velocities that are less than those determined by the EGLA method are subject to the limitation in ASCE/SEI 6.7.6.8. Site-specific velocities greater than those determined by the EGLA method must be used.

Analysis methods that are to be used to determine inundation depths and flow velocities are given in Tables 3.3 and 3.4 based on whether the runup elevation is given in ASCE/SEI Figure 6.1-1 or determined from ASCE/SEI Figure 6.7-1, respectively.

The R/H_T and EGLA methods are covered in the following sections. Information on how to determine maximum inundation depths and flow velocities using a site-specific PTHA is given in ASCE/SEI C6.7.

Analysis Procedure	Tsunami Risk Category			
			IV	
	II	III	Excluding TVERS	TVERS
R/H_T Analysis (ASCE/SEI 6.5.1.1)	Not permitted	Not permitted	Not permitted	Not permitted
Energy Grade Line Analysis (ASCE/SEI 6.6)	Required*	Required*	Required	Required
Site-specific PTHA [ASCE/SEI 6.7]	Permitted*	Permitted*	Required**	Required

*The requirements of ASCE/SEI Chapter 6 do not apply to TRC II and III buildings and other structures where the MCT inundation depth ≤ 3 ft (0.91 m) [ASCE/SEI 6.1.1].

**A site-specific PTHA must be performed where the inundation depth resulting from the EGLA is determined to be greater than or equal to 12 ft (3.7 m) at any point within the location of the TRC IV structure (other than TVERS) [ASCE/SEI 6.5.2].

TABLE 3.3   Inundation Depth and Flow Velocity Analysis Procedures Where Runup Elevation Is Given in ASCE/SEI Figure 6.1-1

Analysis Procedure	Tsunami Risk Category			
			IV	
	II	III	Excluding TVERS	TVERS
R/H_T Analysis (ASCE/SEI 6.5.1.1)	Required	Required	Not permitted	Not permitted
Energy Grade Line Analysis (ASCE/SEI 6.6)	Required*	Required*	Required	Required
Site-specific probabilistic tsunami hazard analysis (PTHA) [ASCE/SEI 6.7]	Permitted*	Permitted*	Required	Required

*The requirements of ASCE/SEI Chapter 6 do not apply to TRC II and III buildings and other structures where the MCT inundation depth ≤ 3 ft (0.91 m) [ASCE/SEI 6.1.1].

TABLE 3.4   Inundation Depth and Flow Velocity Analysis Procedures Where Runup Elevation Is Determined by ASCE/SEI Figure 6.7-1

3.8.2 R/H_T Analysis

As noted in Sec. 3.8.1 of this publication, the runup elevation, R, can be determined using the procedure in ASCE/SEI 6.5.1.1 for TRC II and III buildings and other structures where no mapped inundation limit is shown in ASCE/SEI Figure 6.1-1 (see Table 3.4).

The ratio of tsunami runup elevation to offshore tsunami amplitude, R/H_T, is determined by ASCE/SEI Equations (6.5-2a) through (6.5-2e) as a function of the surf similarity parameter, ξ_{100} (the equations for R/H_T are plotted in ASCE/SEI Figure 6.5-1).

The offshore tsunami amplitude, H_T, is acquired from the ASCE Geodatabase or the local jurisdiction and ξ_{100} is determined by ASCE/SEI Equation (6.5-1):

$$\xi_{100} = \frac{T_{TSU}}{\cot \Phi}\sqrt{\frac{g}{2\pi H_T}} \tag{3.3}$$

In this equation, T_{TSU} is the predominant wave period of the tsunami at the 328-ft (100-m) water depth, which is acquired from the ASCE Geodatabase (the extent of which is given in ASCE/SEI Figure 6.7-1) or the local jurisdiction, and Φ is the mean slope angle of the Nearshore Profile taken from the 328-ft (100-m) water depth to the mean high water (MHW) elevation along the axis of the topographic transect for the site.

The flowchart in Fig. 3.4 can be used to determine R/H_T. Runup elevation, R, for the site is equal to R/H_T times H_T; R can then be used in an EGLA to obtain inundation depths and flow velocities for TRC II and III buildings and other structures.

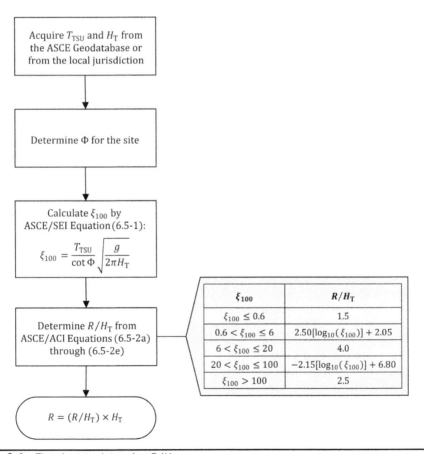

FIGURE 3.4 Flowchart to determine R/H_T.

According to the exception in ASCE/SEI 6.5.1.1, the equations to determine R/H_T must not be used in the following cases:

- Where wave focusing, such as at headlands, are expected;
- In V-shaped bays; or
- Where the on-land flow fields are expected to vary significantly in the direction parallel to the shoreline because of longshore variability of topography.

3.8.3   Energy Grade Line Analysis (EGLA)

The EGLA method is required to determine inundation depths and velocities for the cases identified in Tables 3.3 and 3.4. The main assumption of this conservative, pre-scriptive method is the flow velocity of the tsunami is maximum at the shoreline and zero at the inundation point (ASCE/SEI 6.6.2). Between the shoreline and the inundation point, the flow is decreased by friction from terrain roughness, which is accounted for by Manning's coefficient, n. Energy is accumulated based on the topographic (ground) slope and the equivalent slope due to surface friction. At any location, the total energy is a combination of potential energy (based on the depth of the water, h_i) and kinetic energy (based on the flow velocity, u_i). The Froude number, F_r, which is determined by ASCE/SEI Equation (6.6-3), provides the required relationship between h_i and u_i, so that the hydraulic head, $E_{g,i}$, can be calculated by ASCE/SEI Equation (6.6-1).

The Froude number coefficient, α, is equal to 1.0 where tsunami bores need not be considered. Where tsunami bores must be considered in accordance with ASCE/SEI 6.6.4, tsunami loads on vertical structural components (ASCE/SEI 6.10.2.3) and the hydrodynamic loads for tsunami bore flow entrapped in structural wall-slab recesses (ASCE/SEI 6.10.3.3) must be determined based on inundation depths and flow velocities using $\alpha = 1.3$.

Analysis is performed incrementally across the ground transect at the site from the runup (where the hydraulic head, or energy, at the inundation limit, x_R, is zero) to the shoreline (see Fig. 3.5). The transect must be broken into segments that are no longer than 100 ft (30.5 m) in length; this limit is intended to ensure accuracy of the hydraulic analysis. The ground elevation along the transect is represented by a series of one-dimensional, linear-sloped segments. Normally, the more elevation points that are taken (that is, the smaller the segment lengths), the better the results.

The analysis begins at the runup point, which is located at a horizontal distance x_R from the shoreline, where x_R is obtained from the ASCE Geodatabase or an R/H_T analysis where applicable. At the runup point, the hydraulic head, E_R, is equal to zero. Although the inundation height, h_i, is equal to zero at this point, a small value (such as 0.1 ft or 0.03 m) is used in the analysis because at the next point along the transect, the friction slope, s_i, is determined by ASCE/SEI Equation (6.6-2) or (6.6-2.si) using the inundation depth from the previous point; a value of zero for the inundation depth at the runup point would result in dividing by zero.

The analysis proceeds by calculating the inundation depth, h_i, and the flow velocity, u_i, at every point along the transect; it ends at the shoreline where the hydraulic head and flow velocity are maximum. At the site of the building or other structure, the inundation depth is h_{max} and the flow velocity is u_{max}, which are used to determine tsunami loads and effects on the structure.

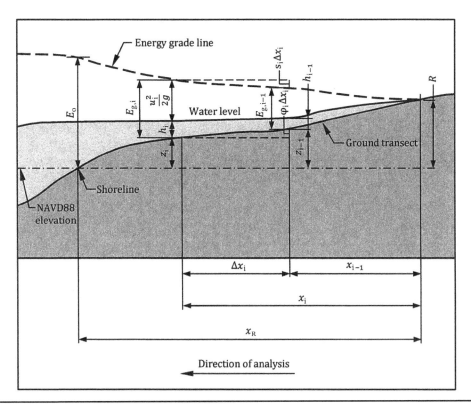

Figure 3.5 Energy Grade Line Analysis (EGLA).

According to ASCE/SEI 6.6.1, the calculated flow velocity must not be taken less than 10 ft/s (3.0 m/s) and need not be taken greater than the lesser of $1.5(gh_{max})^{1/2}$ and 50 ft/s (15.2 m/s). These minimum and maximum limits are based on data from past tsunamis of significant inundation depth. The upper limit of 50 ft/s (15.2 m/s) includes a 1.5 factor on the velocity observed during actual on-land flow.

In cases where the maximum topographic elevation along the transect between the shoreline and the inundation limit is greater than R, one of the following methods must be used to determine h_i and u_i (ASCE/SEI 6.6.1):

1. The site-specific PTHA of ASCE/SEI 6.7.6 subject to the minimum and maximum velocities noted above.

2. The EGLA method of ASCE/SEI 6.6.2 assuming an x_R (and corresponding R) that has at least 100 percent of the maximum topographic elevation along the topographic transect

The flowchart in Fig. 3.6 can be used to determine the inundation depths, h_i, and the flow velocities, u_i, along the transect, including h_{max} and u_{max} at the site of a building or other structure, in accordance with the EGLA method in ASCE/SEI 6.6.2.

The directionality of flow requirements in ASCE/SEI 6.8.6 must be applied in the analysis. Both incoming and outgoing flow conditions must be considered for

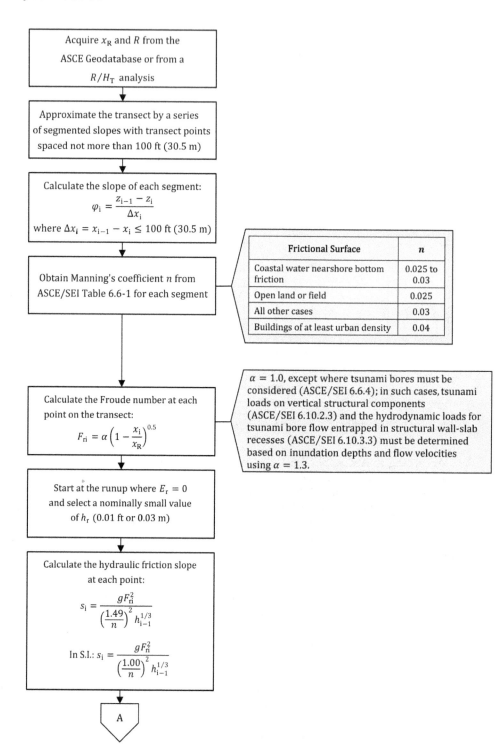

The figure contains the following flowchart text:

Acquire x_R and R from the ASCE Geodatabase or from a R/H_T analysis

Approximate the transect by a series of segmented slopes with transect points spaced not more than 100 ft (30.5 m)

Calculate the slope of each segment:
$$\varphi_i = \frac{z_{i-1} - z_i}{\Delta x_i}$$
where $\Delta x_i = x_{i-1} - x_i \leq 100$ ft (30.5 m)

Obtain Manning's coefficient n from ASCE/SEI Table 6.6-1 for each segment

Frictional Surface	n
Coastal water nearshore bottom friction	0.025 to 0.03
Open land or field	0.025
All other cases	0.03
Buildings of at least urban density	0.04

Calculate the Froude number at each point on the transect:
$$F_{ri} = \alpha \left(1 - \frac{x_i}{x_R} \right)^{0.5}$$

$\alpha = 1.0$, except where tsunami bores must be considered (ASCE/SEI 6.6.4); in such cases, tsunami loads on vertical structural components (ASCE/SEI 6.10.2.3) and the hydrodynamic loads for tsunami bore flow entrapped in structural wall-slab recesses (ASCE/SEI 6.10.3.3) must be determined based on inundation depths and flow velocities using $\alpha = 1.3$.

Start at the runup where $E_r = 0$ and select a nominally small value of h_r (0.01 ft or 0.03 m)

Calculate the hydraulic friction slope at each point:
$$s_i = \frac{gF_{ri}^2}{\left(\frac{1.49}{n}\right)^2 h_{i-1}^{1/3}}$$

In S.I.: $s_i = \frac{gF_{ri}^2}{\left(\frac{1.00}{n}\right)^2 h_{i-1}^{1/3}}$

A

Figure 3.6 Determination of inundation depth and flow velocities using the Energy Grade Line Analysis (EGLA) method.

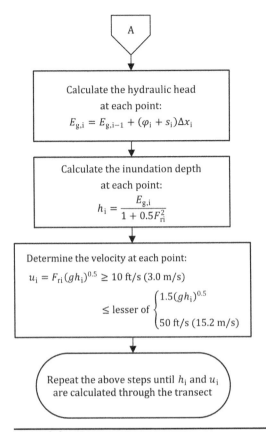

A

Calculate the hydraulic head
at each point:

$$E_{g,i} = E_{g,i-1} + (\varphi_i + s_i)\Delta x_i$$

Calculate the inundation depth
at each point:

$$h_i = \frac{E_{g,i}}{1 + 0.5F_{ri}^2}$$

Determine the velocity at each point:

$$u_i = F_{ri}(gh_i)^{0.5} \geq 10 \text{ ft/s } (3.0 \text{ m/s})$$

$$\leq \text{ lesser of } \begin{cases} 1.5(gh_i)^{0.5} \\ 50 \text{ ft/s } (15.2 \text{ m/s}) \end{cases}$$

Repeat the above steps until h_i and u_i
are calculated through the transect

FIGURE 3.6   *(Continued)*

three flow directions (transects) averaged over 500 ft (152 m) to either side of the site (see Fig. 3.7):

- Transect located perpendicular to the shoreline.
- Transect located 22.5-degree clockwise from the transect that is perpendicular to the shoreline.
- Transect located 22.5-degree counterclockwise from the transect that is perpendicular to the shoreline.

This requirement accounts for tsunamis that do not hit a building head on. The transect producing the maximum runup elevation may be used to perform the EGLA. Alternatively, an EGLA may be performed in each direction, and the one producing the maximum tsunami load effects on the structure should be used.

Flow velocities determined by the EGLA method must be adjusted for flow amplification in accordance with ASCE/SEI 6.8.5, as applicable (ASCE/SEI 6.6.5). The adjusted value need not exceed the maximum flow rate, which, as noted previously, is the lesser of $1.5(gh_{\text{max}})^{0.5}$ and 50 ft/s (15.2 m/s).

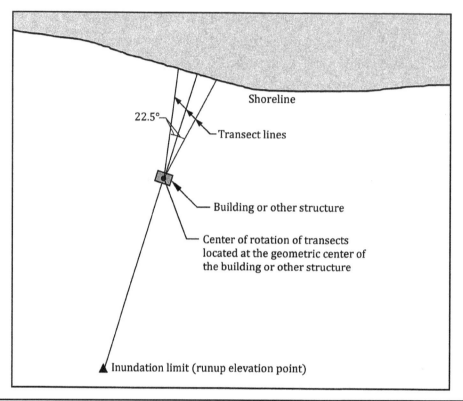

Figure 3.7   Directionality of flow requirements in ASCE/SEI 6.8.6.

3.9 Hydrostatic Loads

3.9.1 Buoyancy Loads

Buoyancy loads, F_v, must be evaluated for all inundated members of a building or other structure by ASCE/SEI Equation (6.9-1):

$$F_v = \gamma_s V_w \tag{3.4}$$

where γ_s = minimum fluid weight density for design hydrostatic loads

$\quad\quad\quad$ = $k_s \gamma_{sw}$ [ASCE/SEI Equation (6.8-4)]

$\quad\quad k_s$ = fluid density factor = 1.1

$\quad\quad \gamma_{sw}$ = specific weight of seawater = 64.0 lb/ft³ (10.05 kN/m³)

$\quad\quad V_w$ = displaced water volume

The buoyancy load on a building with a watertight exterior is illustrated in Fig. 3.8 where h_{max} is the maximum inundation depth at the structure. This load is resisted by the weight of the building and the foundation (the weight of any deep foundations must not be included to resist uplift loads).

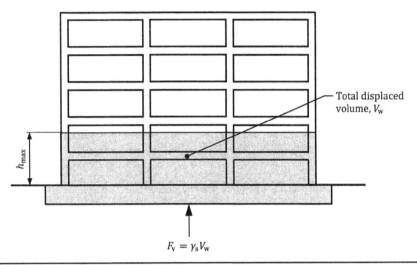

Total displaced
volume, V_w

$F_v = \gamma_s V_w$

FIGURE 3.8 Buoyancy load on a building with a watertight exterior.

For purposes of discussion, a watertight exterior is assumed to be designed and detailed to resist all load effects from tsunamis, which means that the exterior will not be breached. Unless designed for large missile wind-borne debris impact or blast loading, windows must be considered as openings when the inundation depth reaches the top of the windows or the expected strength of the glazing (whichever is less), which means water can flow into enclosed spaces, thereby decreasing or eliminating buoyancy loads (ASCE/SEI 6.9.1).

Buoyancy pressures on slabs in buildings without and with a basement are illustrated in Fig. 2.8 of this publication where the buildings have a watertight exterior. In cases where the slab is isolated from the building structure (which is a typical slab-on-grade), the slab is subjected to the buoyancy pressure and overall buoyancy (uplift) on the building need not be considered. However, if the slab is connected to the building structure and is designed to resist the buoyancy pressure, uplift on the building must be considered.

According to ASCE/SEI 6.9.1, uplift caused by buoyancy must be considered for the following:

- Enclosed spaces without tsunami breakaway walls where the walls have opening areas less than 25 percent of the inundated exterior wall area.
- Floors with trapped air below, including integral structural slabs and in enclosed spaces where the walls are not designed to break away during a tsunami event.

3.9.2 Unbalanced Lateral Hydrostatic Load

Lateral hydrostatic loads develop where there is a difference in water level on either side of a member. Such loads can develop on walls with relatively long widths (water tends to equalize on both sides of walls with shorter widths) or on walls in the lower levels of a watertight building.

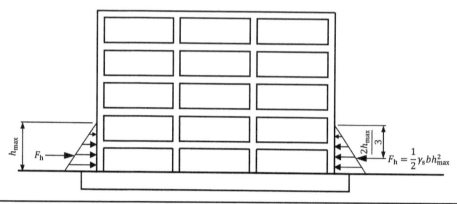

Figure 3.9 Unbalanced lateral hydrostatic load.

Unbalanced hydrostatic loads must be considered on the following inundated walls with openings less than 10 percent of the wall area (ASCE/SEI 6.9.2):

- Walls longer than 30 ft (9.1 m) in width without adjacent tsunami breakaway walls

- Two- or three-sided walls of any width where water is prevented from getting to the other side of the wall

ASCE/SEI Equation (6.9-2) is used to calculate unbalanced hydrostatic loads, F_h, where required:

$$F_h = \frac{1}{2}\gamma_s b h_{max}^2 \tag{3.5}$$

where b is the width of the wall. Illustrated in Fig. 3.9 is the unbalanced lateral hydrostatic load on a building with exterior walls that satisfy the criteria in ASCE/SEI 6.9.2.

3.9.3 Residual Water Surcharge Load on Floors and Walls

Where water is unable to drain on floors located below h_{max} in buildings that are not watertight, the floors must be designed to resist the following residual water surcharge pressure, p_r, in addition to all other applicable loads [ASCE/SEI Equation (6.9-3)]:

$$p_r = \gamma_s h_r = \gamma_s (h_{max} - h_s) \tag{3.6}$$

where h_r is the residual water height in the building, which is equal to h_{max} minus the height of the structural floor slab above grade, h_s. Note that h_r need not exceed the height of the continuous portion of any perimeter structural element at the floor. This is illustrated in Fig. 3.10 for a continuous upturned reinforced concrete beam at the perimeter of the floor located below h_{max} in a building that is not watertight. The floor slab is subjected to p_r determined in accordance with Eq. (3.6) where h_r is the depth of the upturned beam above the floor slab. Also, the upturned beam is subjected to lateral hydrostatic pressure caused by the residual water, which is equal to zero at the top of the upturned beam and varies linearly over its depth with a maximum pressure of $\gamma_s h_r$ at the top of the floor slab (see Fig. 3.10).

Nonstructural elements above the perimeter structural elements are assumed to fail during tsunami inflow, which means they do not contribute to the retention of water on the slab during drawdown.

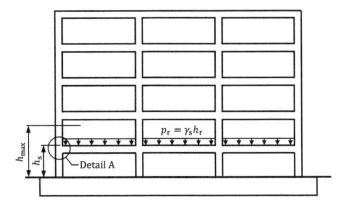

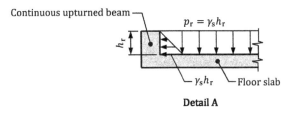

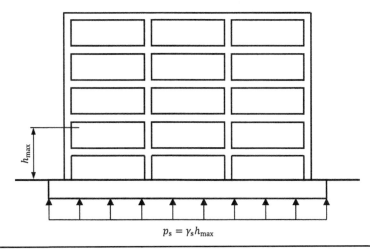

FIGURE 3.10 Determination of residual water pressure, p_r.

3.9.4 Hydrostatic Surcharge Pressure on Foundations

Different water levels may exist on opposite sides of a building, wall, or other structure during tsunami inundation and drawdown. The resulting differential in hydrostatic pressure on the foundation, p_s, that can occur is determined by ASCE/SEI Equation (6.9-4) [see Fig. 3.11]:

$$p_s = \gamma_s h_{max} \qquad (3.7)$$

FIGURE 3.11 Hydrostatic surcharge pressure due to tsunami inundation on a foundation.

3.10 Hydrodynamic Loads

3.10.1 Overview
Hydrodynamic loads, which are often referred to as drag forces, develop when water flows around an object in the flow path. According to ASCE/SEI 6.10, the structure's lateral force–resisting system and all structural components below the inundation level at the site must be designed for the hydrodynamic loads in accordance with the simplified method in ASCE/SEI 6.10.1 or the detailed method in ASCE/SEI 6.10.2.

3.10.2 Simplified Equivalent Uniform Lateral Static Pressure
In lieu of performing a more detailed analysis, ASCE/SEI Equation (6.10-1) is permitted to be used to determine an equivalent maximum uniform pressure, p_{uw}, which accounts for the combination of any unbalanced lateral hydrostatic and hydrodynamic loads caused by a tsunami:

$$p_{uw} = 1.25 I_{tsu} \gamma_s h_{max} \tag{3.8}$$

where I_{tsu} is the importance factor for tsunami forces given in ASCE/SEI Table 6.8-1.

This pressure is to be applied uniformly over a depth of $1.3h_{max}$ at the site in each flow direction (see Fig. 3.12).

Equation (3.8) is based on the assumption that the most conservative requirements in ASCE/SEI 6.10 occur at the same time on a rectangular building with no openings (see ASCE/SEI C6.10.1 for the derivation of this equation). The lateral force–resisting system is evaluated for p_{uw} acting over the entire width of the building perpendicular to the flow direction for both incoming and outgoing flow, and all structural members below $1.3h_{max}$ must be evaluated for the effects of this pressure on their tributary width of projected area.

3.10.3 Detailed Hydrodynamic Lateral Forces
Drag forces on the entire building and on components must be considered for incoming and outflowing flow. Both of these cases are examined below.

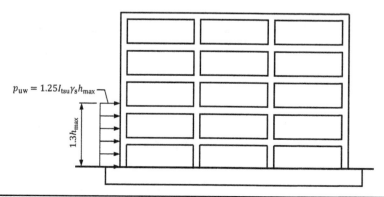

$p_{uw} = 1.25 I_{tsu} \gamma_s h_{max}$

$1.3h_{max}$

FIGURE 3.12 Simplified equivalent uniform lateral static pressure.

Overall Drag Force on Buildings and Other Structures

The overall drag force on a building or other structure, F_{dx}, is determined by ASCE/SEI Equation. (6.10-2) at each level caused by either incoming or outgoing flow for load case 2 defined in ASCE/SEI 6.8.3.1 (see Sec. 3.7.1 of this publication):

$$F_{dx} = \frac{1}{2}\rho_s I_{tsu} C_d C_{cx} B(hu^2) \tag{3.9}$$

In this equation, h and u are the inundation depths and flow velocities, respectively, determined at the site for load case 2, that is, $h = 2h_{max}/3$ and $u = u_{max}$.

The flowchart in Fig. 3.13 can be used to determine F_{dx}. This load is applied to the lateral force–resisting system of the building at each inundated level.

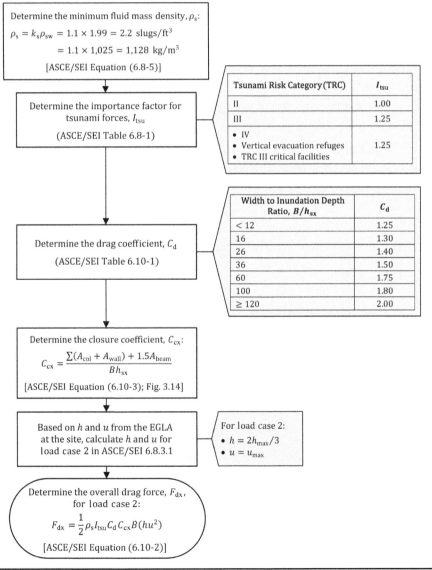

FIGURE 3.13 Flowchart to determine the overall drag force on a building or other structure, F_{dx}.

The closure coefficient, C_{cx}, which is determined by ACI Equation (6.10-3), is the percentage of the total projected area of the inundated portion of the building that can block floating debris; that is, this coefficient accounts for debris accumulation during the tsunami event. The numerator in the equation represents the vertical projected area of inundated structural components (columns, walls, and beams, as applicable) and the denominator represents the vertical projected area of the submerged section of the building. Identified in Fig. 3.14 are the projected areas of the beams (A_{beam}), columns (A_{col}), and walls (A_{wall}) for the building in the figure. The summation of A_{col} and A_{wall} is over all the columns and walls in the building, and A_{beam} is the combined vertical projected area of the slab facing the flow and the deepest beam laterally exposed to the flow. The equation for C_{cx} in the direction of flow indicated for this building is given in Fig. 3.14.

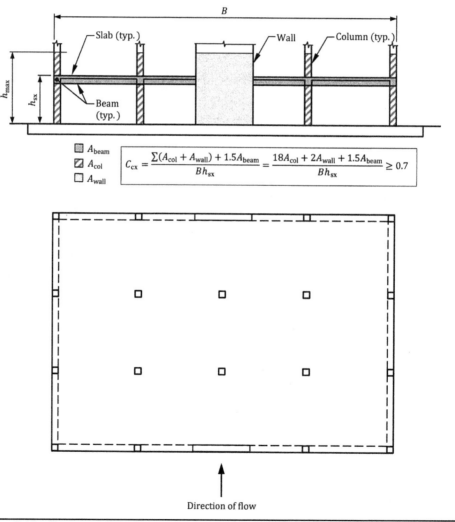

$$C_{cx} = \frac{\Sigma(A_{col} + A_{wall}) + 1.5A_{beam}}{Bh_{sx}} = \frac{18A_{col} + 2A_{wall} + 1.5A_{beam}}{Bh_{sx}} \geq 0.7$$

FIGURE 3.14   Closure coefficient, C_{cx}.

According to ASCE/SEI 6.8.7, C_{cx} must be at least 0.7 when determining F_{dx} by ASCE/SEI Equation (6.10-2).

Drag Force on Components

All structural components and wall assemblies of the building below the inundation depth are subject to the component drag force, F_d, which is determined by ASCE/SEI Equation (6.10-4):

$$F_d = \frac{1}{2}\rho_s I_{tsu} C_d b (h_e u^2) \qquad (3.10)$$

where h_e is the inundation height of an individual element.

The flowchart in Fig. 3.15 can be used to determine F_d. This load is applied as a distributed load on the submerged portion of the component and need not be added to the overall drag force calculated in accordance with ASCE/SEI 6.10.2.1.

For interior components, C_d is determined by ASCE/SEI Table 6.10-2 based on the section of the component and b is the width of the component perpendicular to the flow. The effects of debris accumulation are considered for exterior components: $C_d = 2.0$ and b is taken as the tributary width of the component times C_{cx} given in ASCE/SEI 6.8.7, which is equal to 0.7 for other than open structures (see ASCE/SEI 6.2 for the definition of an open structure).

Tsunami Loads on Vertical Structural Components

The hydrodynamic drag force, F_w, on a vertical component, such as wall, that is wider than 3 times the inundation depth corresponding to load case 2 in ASCE/SEI 6.8.3.1 during inflow where flow of a tsunami bore need not be considered is determined by ASCE/SEI Equation (6.10-5a); that equation is the same as ASCE/SEI Equation (6.10-4).

ASCE/SEI Equation (6.10-5b) must be used to determine F_w where all the following additional conditions are satisfied:

- Flow of a tsunami bore must be considered.
- The Froude number at the site is greater than 1.0.
- Individual wall, wall pier, or column components have a width to inundation depth ratio of 3 or more.

The force F_w determined by ASCE/SEI Equation (6.10-5b) is equal to 1.5 times the force F_w determined by ASCE/SEI Equation (6.10-5a).

A more detailed method to determine the force when a bore strikes a wall is given in ASCE/SEI C6.10.2.3.

Hydrodynamic Load on Perforated Walls

For walls with openings that permit flow to pass between wall piers, the force on the elements of the perforated wall, F_{pw}, is determined by ASCE/SEI (6.10-6) but must not be taken less than F_d determined by ASCE/SEI Equation (6.10-4):

$$F_{pw} = (0.4C_{cx} + 0.6)F_w \ge F_d = \frac{1}{2}\rho_s I_{tsu} C_d b(h_e u^2) \qquad (3.11)$$

where F_w is determined by ASCE/SEI Equation (6.10-5a) or (6.10-5b).

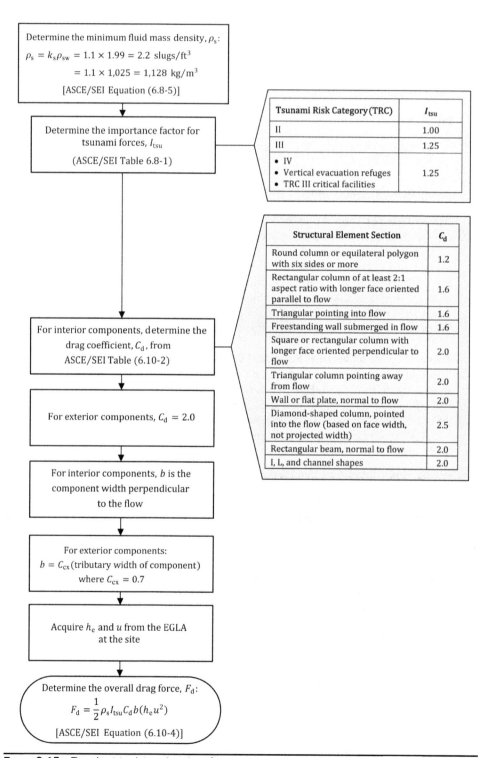

Figure 3.15 Flowchart to determine drag force on components, F_d.

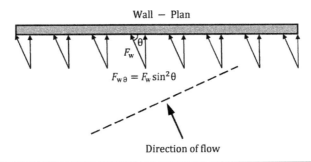

Wall — Plan

F_w

$F_{w\theta} = F_w \sin^2\theta$

Direction of flow

FIGURE 3.16 Lateral hydrodynamic load on a wall angled to the flow.

Walls Angled to the Flow

ASCE/SEI Equation (6.10-7) is used to calculate the lateral load per unit width, $F_{w\theta}$, for walls oriented at an angle of less than 90 degrees to the flow direction (see Fig. 3.16):

$$F_{w\theta} = F_w \sin^2\theta \qquad (3.12)$$

where θ is the angle between the wall and the direction of the flow and F_w is determined by ASCE/SEI Equation (6.10-5a) or (6.10-5b).

A summary of the detailed hydrodynamic lateral forces is given in Table 3.5.

	Hydrodynamic Lateral Force	ASCE/SEI Equation No.	Notes
Overall drag force on buildings and other structures	$F_{dx} = \frac{1}{2}\rho_s I_{tsu} C_d C_{cx} B(hu^2)$	6.10-2	• See Fig. 3.13 for determination of F_{dx} • See Fig. 3.14 for determination of C_{cx}
Drag forces on components	$F_d = \frac{1}{2}\rho_s I_{tsu} C_d b(h_e u^2)$	6.10-4	• See Fig. 3.15 for determination of F_d
Tsunami loads on vertical structural components	• Where bores do not need to be considered: $F_w = \frac{1}{2}\rho_s I_{tsu} C_d b(h_e u^2)$ • Where bores must be considered along with the other conditions in ASCE/SEI 6.10.2.3: $F_w = \frac{3}{4}\rho_s I_{tsu} C_d b(h_e u^2)_{bore}$	• 6.10-5a • 6.10-5b	• See Fig. 3.15 for determination of F_w where bores do not need to be considered
Hydrodynamic loads on perforated walls	$F_{pw} = (0.4C_{cx} + 0.6)F_w \geq F_d$	6.10-6	• See Fig. 3.15 for determination of F_w
Walls angled to the flow	$F_{w\theta} = F_w \sin^2\theta$	6.10-7	• See Fig. 3.15 for determination of F_w • See Fig. 3.16

TABLE 3.5 Summary of Detailed Hydrodynamic Lateral Forces in ASCE/SEI 6.10.2

3.10.4 Hydrodynamic Pressures Associated with Slabs

A summary of the requirements in ASCE/SEI 6.10.3 pertaining to hydrodynamic pressures associated with slabs is given in Table 3.6.

Flow stagnation pressure	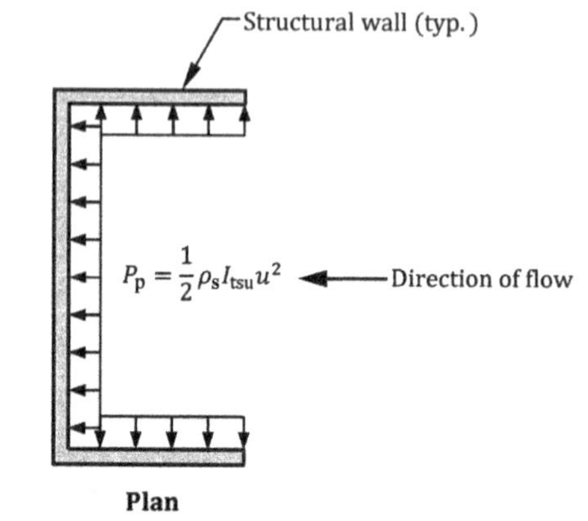Structural wall (typ.) $P_p = \frac{1}{2}\rho_s I_{tsu} u^2$ Direction of flow Plan
Slabs over sloping grade	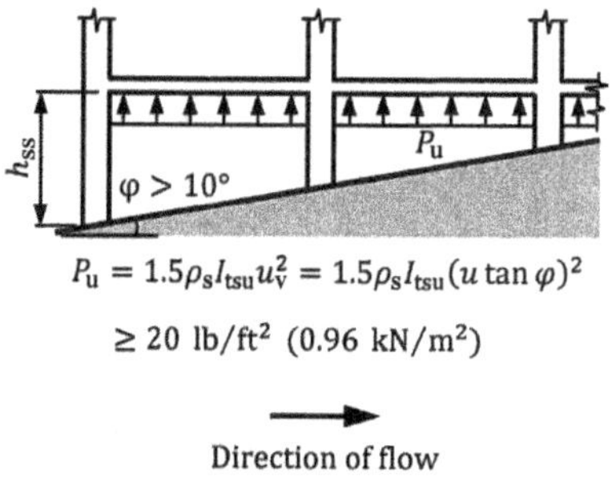h_{ss} P_u $\varphi > 10°$ $P_u = 1.5\rho_s I_{tsu} u_v^2 = 1.5\rho_s I_{tsu}(u \tan\varphi)^2$ ≥ 20 lb/ft² (0.96 kN/m²) Direction of flow
Tsunami bore flow entrapped in structural wall-slab recess	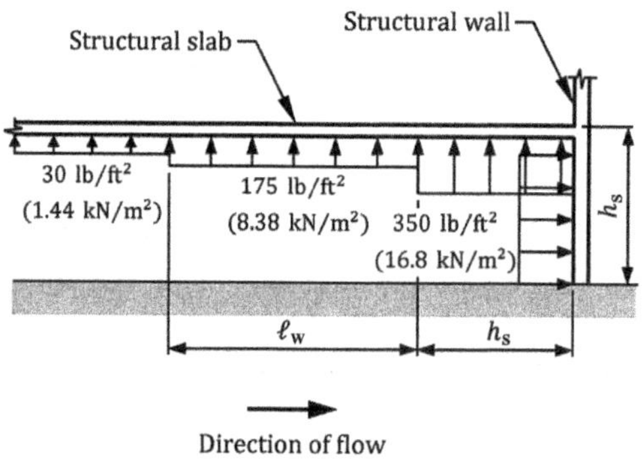Structural slab Structural wall 30 lb/ft² (1.44 kN/m²) 175 lb/ft² (8.38 kN/m²) 350 lb/ft² (16.8 kN/m²) h_s ℓ_w h_s Direction of flow

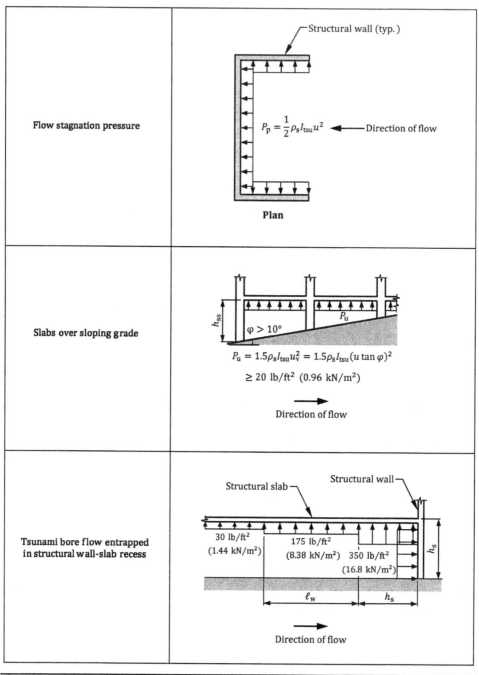

Table 3.6 Hydrodynamic Pressures Associated with Slabs

Flow Stagnation Pressure

Enclosed spaces created by structural walls on three sides and structural slabs can become pressurized by flow entering the wall-enclosed space with no openings. Therefore, walls and slabs in such spaces must be subjected to the following pressure [ASCE/SEI Equation (6.10-8)]:

$$P_p = \frac{1}{2}\rho_s I_{tsu} u^2 \tag{3.13}$$

where u is the maximum free flow velocity at that location and load case in accordance with ASCE/SEI 6.8.3.1.

Hydrodynamic Surge Uplift at Horizontal Slabs

Uplift pressures can occur during tsunami inflow on submerged horizontal slabs with no flow obstructions (such as walls or columns) above or below the slab. A 20 lb/ft² (0.96 kN/m²) uplift pressure must be applied to the soffit of the slab in such cases (ASCE/SEI 6.10.3.2.1). This pressure is in addition to any buoyancy effects required by ASCE/SEI 6.9.1. Significantly larger uplift pressures occur where obstructions are present below the slab (see below).

Horizontal elevated slabs located over sloping ground are subjected to upward hydrodynamic pressure when the flow reaches the elevation of the slab soffit. Where the ground has an average slope, φ, greater than 10 degrees, the uplift pressure, P_u, is determined by ASCE/SEI Equation (6.10-9):

$$P_u = 1.5 I_{tsu}\rho_s u_v^2 \geq 20 \text{ lb/ft}^2 \text{ (0.96 kN/m}^2) \tag{3.14}$$

where $u_v = u\tan\varphi$ is the vertical component of the flow velocity and u is the horizontal flow velocity corresponding to a water depth equal to or greater than the elevation to the soffit, h_{ss}.

Tsunami Bore Flow Entrapped in Structural Wall-Slab Recesses

Significant uplift pressure can occur below horizontal structural slabs where the flow below the slab is blocked by a wall (or other obstruction); in such cases, flow is diverted upward to the underside of the slab. Large pressures can occur on the wall and the slab soffit close to the face of the wall.

The wall and slab within a distance of h_s from the face of the wall must be subjected to an outward pressure, P_u, equal to 350 lb/ft² (16.8 kN/m²) where h_s is the height of the structural floor slab above the grade plane (ASCE/SEI 6.10.3.3.1). Between h_s and the length of the wall, ℓ_w, the upward pressure on the slab is equal to 175 lb/ft² (8.38 kN/m²). Beyond ℓ_w, the required upward pressure is 30 lb/ft² (1.44 kN/m²).

Uplift pressures specified in ASCE/SEI 6.10.3.3.1 may be reduced based on inundation depth (ASCE/SEI 6.10.3.3.2), wall openings (ASCE/SEI 6.10.3.3.3), slab openings (ASCE/SEI 6.10.3.3.4), and tsunami breakaway walls (ASCE/SEI 6.10.3.3.5).

3.11 Debris Impact Loads

3.11.1 Overview

Large volumes of debris are typically transported by tsunamis, ranging in size from logs to ships. Requirements for debris impact loads are given in ASCE/SEI 6.11. All buildings and other structures located where the minimum inundation depth is 3 ft (0.91 m) or greater must be designed for debris impact loads due to the following:

- Floating wood logs and poles (ASCE/SEI 6.11.2)
- Vehicles (ASCE/SEI 6.11.3)
- Tumbling boulders and concrete debris (ASCE/SEI 6.11.4)

For buildings or other structures located in proximity to a port or container year, debris impact loads from shipping containers, ships, and barges may also be applicable (ASCE/SEI 6.11.5 through 6.11.7).

In general, debris impact loads must be applied at points critical for flexure and shear on all structural elements supporting gravity loads at the perimeter of the building, except as specified otherwise in the applicable provisions. Inundation depths and flow velocities corresponding to load cases 1, 2, and 3 in ASCE/SEI 6.8.3.1 must be used.

An alternative simplified debris impact static load method is given in ASCE/SEI 6.11.1, which is permitted to be used to evaluate debris impact loads instead of the methods noted above.

Debris impact loads are considered to be a separate load case, which means that they do not have to be applied simultaneously to all affected structural components (ASCE/SEI 6.11).

A summary of the conditions for which debris impact loads must be evaluated is given in Table 3.7 (see ASCE/SEI Table C6.11-1).

A summary of the debris impact loads in ASCE/SEI 6.11.2 through 6.11.7 is given in Table 3.8.

In lieu of the prescriptive methods in Table 3.8, a dynamic analysis is permitted to be used to determine debris impact loads (ASCE/SEI 6.11.8). Information on how to perform such an analysis is given in ASCE/SEI C6.11.8.

3.11.2 Alternative Simplified Debris Impact Static Load

It is permitted to determine the debris impact load, F_i, by ASCE/SEI Equations (6.11-1) and (6.11-1.si) instead of the detailed methods in ASCE/SEI 6.11.2 through 6.11.6:

$$F_i = 330C_o I_{tsu} \text{ (kips)} \tag{3.15a}$$

$$F_i = 1,470C_o I_{tsu} \text{ (kN)} \tag{3.15b}$$

In these equations, C_o is the orientation factor, which is equal to 0.65.

These equations, which are derived in ASCE/SEI C6.11.1, give conservative values of impact loads for poles, logs, vehicles, tumbling boulders, concrete debris, and shipping containers. The load is applied at points critical for flexure and shear on the perimeter structural members in the inundation depth corresponding to load case 3 in ASCE/SEI 6.8.3.1.

It is permitted to reduce the calculated value of F_i by 50 percent for buildings or other structures not located in an impact zone for shipping containers, ships, and barges, which is determined by the hazard assessment procedure in ASCE/SEI 6.11.5.

3.11.3 Wood Logs and Poles

The flowchart in Fig. 3.17 can be used to determine debris impact loads due to wood logs and poles in accordance with ASCE/SEI 6.11.2. It is assumed that a log or pole hits the structural element longitudinally rather than transversely. In lieu of a dynamic analysis,

Debris	Buildings and Other Structures[1]	Threshold Inundation Depth, ft (m)
Poles, logs, and passenger vehicles	All	3 (0.91)
Boulders and concrete debris	All	6 (1.8)
Shipping containers	All	3 (0.91)
Ships and/or barges[2]	TRC III critical facilities TRC IV	12 (3.7)

[1]"All" refers to buildings and other structures located within a TDZ (ASCE/SEI 6.1.1).
[2]Applicable to TRC III critical facilities and TRC IV buildings and other structures located in the debris hazard impact region determined in accordance with ASCE/SEI 6.11.5.

TABLE 3.7 Conditions for Which Design for Debris Impact Is Evaluated

Debris	Debris Impact Load	Notes
Wood logs and poles (ASCE/SEI 6.11.2)	• Nominal maximum instantaneous debris impact load: $$F_{ni} = u_{max}\sqrt{km_d}$$ • Design instantaneous debris impact load: $$F_i = I_{tsu}C_oF_{ni}$$ • Equivalent elastic static analysis: $$\text{Impact force} = R_{max}F_i$$	See Fig. 3.17 for determination of impact load.
Vehicles (ASCE/SEI 6.11.3)	$F_i = 30I_{tsu}$ (kips) $F_i = 130I_{tsu}$ (kN)	F_i is applied to the vertical structural elements at any point greater than 3 ft (0.91 m) above grade up to the maximum depth.
Tumbling boulder and concrete debris (ASCE/SEI 6.11.4)	$F_i = 8I_{tsu}$ (kips) $F_i = 36I_{tsu}$ (kN)	F_i is applied to the vertical structural elements at 2 ft (0.61 m) above grade.
Shipping containers (ASCE/SEI 6.11.6)	• Nominal maximum instantaneous debris impact load: $$F_{ni} = u_{max}\sqrt{km_d} \leq 220 \text{ kips (980 kN)}$$ • Design instantaneous debris impact load: $$F_i = I_{tsu}C_oF_{ni}$$ • Equivalent elastic static analysis: $$\text{Impact force} = R_{max}F_i$$	See Fig. 3.18 for determination of impact load.
Ships and/or barges (ASCE/SEI 6.11.7)	• Nominal maximum instantaneous debris impact load: $$F_{ni} = u_{max}\sqrt{km_d}$$ • Design instantaneous debris impact load: $$F_i = I_{tsu}C_oF_{ni}$$	• See Fig. 3.19 for determination of impact load. • Impact load is applied anywhere from the base of the structure up to 1.3 times the inundation depth plus the height to the deck of the vessel.

TABLE 3.8 Debris Impact Loads

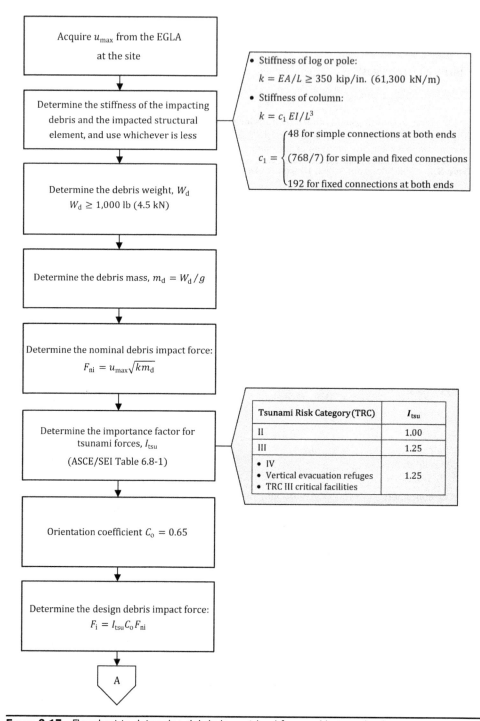

Figure 3.17 Flowchart to determine debris impact load for wood logs and poles.

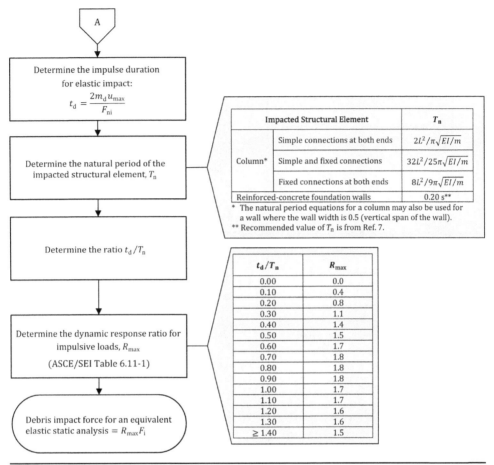

The flowchart begins at connector **A**, leading to:

Determine the impulse duration for elastic impact:

$$t_d = \frac{2m_d u_{max}}{F_{ni}}$$

Determine the natural period of the impacted structural element, T_n

Impacted Structural Element		T_n
Column*	Simple connections at both ends	$2L^2/\pi\sqrt{EI/m}$
	Simple and fixed connections	$32L^2/25\pi\sqrt{EI/m}$
	Fixed connections at both ends	$8L^2/9\pi\sqrt{EI/m}$
Reinforced-concrete foundation walls		0.20 s**

* The natural period equations for a column may also be used for a wall where the wall width is 0.5 (vertical span of the wall).
** Recommended value of T_n is from Ref. 7.

Determine the ratio t_d/T_n

Determine the dynamic response ratio for impulsive loads, R_{max}
(ASCE/SEI Table 6.11-1)

t_d/T_n	R_{max}
0.00	0.0
0.10	0.4
0.20	0.8
0.30	1.1
0.40	1.4
0.50	1.5
0.60	1.7
0.70	1.8
0.80	1.8
0.90	1.8
1.00	1.7
1.10	1.7
1.20	1.6
1.30	1.6
≥ 1.40	1.5

Debris impact force for an equivalent elastic static analysis = $R_{max}F_i$

FIGURE 3.17 *(Continued)*

an equivalent elastic static analysis is permitted where the dynamic impact force is obtained by multiplying the impact force, F_i, by the dynamic response ratio for impulsive loads, R_{max}, given in ASCE/SEI Table 6.11-1. The dynamic response ratio is a function of the ratio of the impact duration, t_d, determined by ASCE/SEI Equation (6.11-4), to the natural period of the impacted structural element, T_n. Approximate equations to determine T_n for columns with various end conditions are given in Fig. 3.17.

For walls, impact is assumed to act along the horizontal center of the wall. The natural period of the wall is permitted to be determined based on the fundamental period of an equivalent column with a width equal to one-half the vertical span of the wall.

3.11.4 Impact by Vehicles

Requirements for debris impact loads due to vehicles are given in ASCE/SEI 6.11.3. Passenger vehicles easily float during a tsunami event. The impact load, F_i, in kips is equal to $30I_{tsu}$ ($130I_{tsu}$ in kN). The derivation of the 30-kip (130-kN) impact force is given

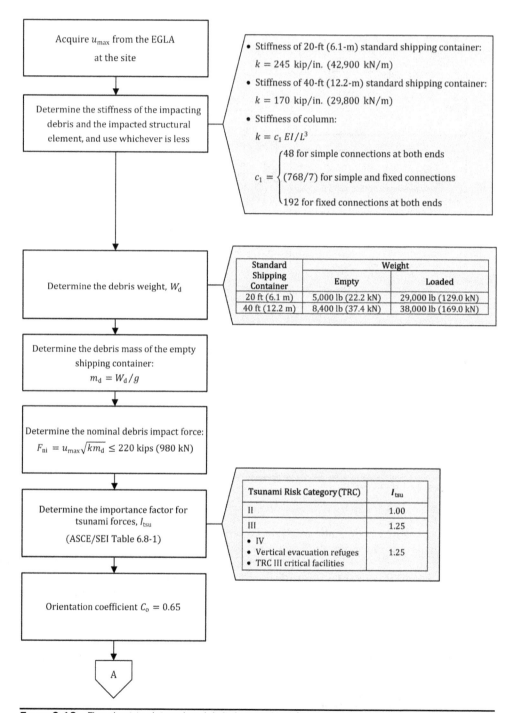

Figure 3.18 Flowchart to determine debris impact load for shipping containers.

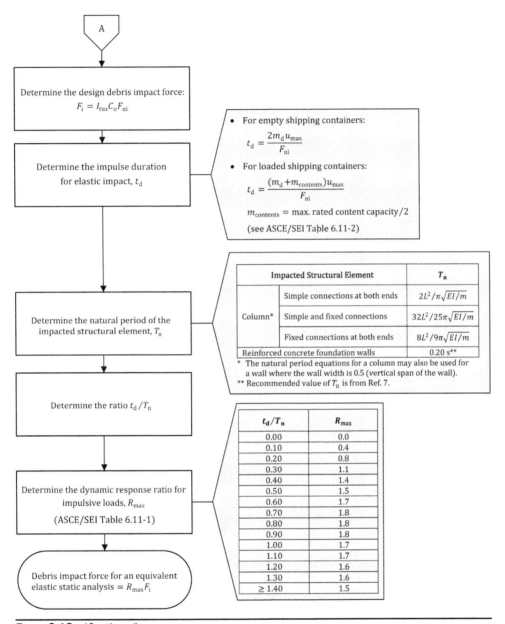

Determine the design debris impact force:
$$F_i = I_{tsu} C_o F_{ni}$$

Determine the impulse duration for elastic impact, t_d

- For empty shipping containers:
$$t_d = \frac{2 m_d u_{max}}{F_{ni}}$$

- For loaded shipping containers:
$$t_d = \frac{(m_d + m_{contents}) u_{max}}{F_{ni}}$$

$m_{contents}$ = max. rated content capacity/2

(see ASCE/SEI Table 6.11-2)

Determine the natural period of the impacted structural element, T_n

Impacted Structural Element		T_n
Column*	Simple connections at both ends	$2L^2/\pi \sqrt{EI/m}$
	Simple and fixed connections	$32L^2/25\pi \sqrt{EI/m}$
	Fixed connections at both ends	$8L^2/9\pi \sqrt{EI/m}$
Reinforced concrete foundation walls		0.20 s**

* The natural period equations for a column may also be used for a wall where the wall width is 0.5 (vertical span of the wall).
** Recommended value of T_n is from Ref. 7.

Determine the ratio t_d/T_n

Determine the dynamic response ratio for impulsive loads, R_{max}

(ASCE/SEI Table 6.11-1)

t_d/T_n	R_{max}
0.00	0.0
0.10	0.4
0.20	0.8
0.30	1.1
0.40	1.4
0.50	1.5
0.60	1.7
0.70	1.8
0.80	1.8
0.90	1.8
1.00	1.7
1.10	1.7
1.20	1.6
1.30	1.6
≥ 1.40	1.5

Debris impact force for an equivalent elastic static analysis = $R_{max} F_i$

Figure 3.18   *(Continued)*

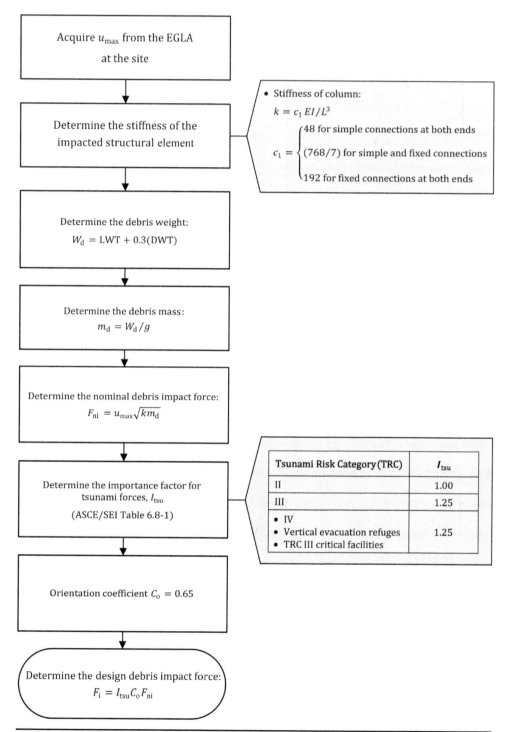

FIGURE 3.19 Flowchart to determine debris impact load for ships and/or barges (extraordinary debris impacts).

in ASCE/SEI C6.11.3. This load is applied to vertical structural elements at any point greater than 3 ft (0.91 m) above grade up to the maximum inundation depth.

3.11.5   Impact by Submerged Tumbling Boulders and Concrete Debris

For submerged tumbling boulders and concrete debris, the impact load, F_i, in pounds is equal to $8,000I_{tsu}$ ($36I_{tsu}$ in kN) [ASCE/SEI 6.11.4]. The derivation of the 8,000-lb (36-kN) impact force is given in ASCE/SEI C6.11.4. It is applied to vertical structural elements at 2 ft (0.61 m) above grade.

3.11.6   Shipping Containers

Debris impact loads due to shipping containers must be considered for any building or other structure located within a debris impact hazard region, which can occur near a port or container yard. Information on how to determine such a region is given in ASCE/SEI 6.11.5 (see ASCE/SEI Figure 6.11-1).

   The flowchart in Fig. 3.18 can be used to determine the debris impact loads due to shipping containers (ASCE/SEI 6.11.6). The mass, m_d, used in determining the nominal maximum instantaneous debris impact force, F_{ni}, by ASCE/SEI Equation (6.11-2) is the mass of the empty shipping container, which is equal to W_d/g. The term W_d is the weight of the empty shipping container given in ASCE/SEI Table 6.11-2 for 20-ft (6.1-m) and 40-ft (12.2-m) standard shipping containers.

   Design impact forces must be determined for both empty and loaded shipping containers. The impulse duration for elastic impact, t_d, which is used in determining the dynamic response ratio for impulsive loads, R_{max}, is calculated by ASCE/SEI Equation (6.11-4) for empty shipping containers. ASCE/SEI Equation (6.11-5) is used to determine t_d for loaded shipping containers where $(m_d + m_{contents})$ is equal to the loaded shipping container weights in ASCE/SEI Table 6.11-2 divided by g.

3.11.7   Extraordinary Debris Impacts

Impact by ships and/or barges is considered an extraordinary event, and must be considered for TRC III Critical Facilities and TRC IV buildings and other structures located in the debris hazard impact region of a port or harbor as defined in ASCE/SEI 6.11.5. Typical vessel sizes may be acquired from the local harbormaster or port authority, or may be obtained in Appendix C of Ref. 11.

3.12   Tsunami Flow Cycles

According to ASCE/SEI 6.8.8, a minimum of two tsunami inflow and outflow cycles must be considered in design:

- The first cycle is based on an inundation depth at 80 percent of the MCT.
- The second cycle is assumed to occur with the MCT inundation depth at the site.

   The purpose of this requirement is to account for any changes that may occur to the condition of the structure and its foundation in each inflow and outflow cycle. Local scour effects determined by the first cycle must be determined in accordance with ASCE/SEI 6.12 based on an inundation depth equal to 80 percent of the MCT design level. The second tsunami cycle is at the MCT design level where the scour from the first cycle is combined with the load effects generated by the inflow of the second cycle.

3.13 Examples

The following examples illustrate the determination of tsunami loads in various TDZs. The steps in Fig. 3.1 are used to determine F_{TSU}.

Although the latitude and longitude of the building sites are given in the examples, the assumed ground elevations along the transects have not been obtained using a topographic digital elevation model (DEM) of at least 33-ft (10-m) resolution, which is recommended in ASCE/SEI C6.6.2 when performing an EGLA at a site.

3.13.1 Example 3.1—Seven-Story Essential Facility

Determine the tsunami loads, F_{TSU}, on the seven-story essential facility in Fig. 3.20. Assume the following:

- Building site:
  - Latitude: 36.59989°
  - Longitude: −121.89101°
- The building is located 820 ft (249.9 m) from the shoreline as shown in Fig. 3.21.
- The ground transect perpendicular to the shoreline is given in Fig. 3.21.
- Runup elevation = 29.6 ft (9.0 m) located 1,570 ft (478.5 m) from the shoreline (obtained from the ASCE Geodatabase).
- No tsunami bores are expected within this area.
- The site is located in an impact zone for shipping containers in accordance with ASCE/SEI 6.11.5.
- Windows are not designed for large impact loads or blast loads and span from the top of the slab to the underside of the beams.
- The tops of the beams are flush with the top of the slab.
- The structural members are constructed of reinforced concrete with a unit weight equal to 150 lb/ft³ (23.6 kN/m³) and a modulus of elasticity, E, equal to 3,605,000 lb/in.² (24,900 MPa).

Solution

Step 1—Determine the Tsunami Risk Category   Table 3.2

From the design data, the building is an essential facility, so the TRC is IV.

Step 2—Determine the permitted analysis procedure   Sec. 3.8

Because the runup elevation is obtained from the ASCE Geodatabase, the permitted analysis procedures are given in Table 3.3.

For a TRC IV building, the EGLA method and a site-specific PTHA must both be performed. It is determined in Step 4 below from an EGLA that $h_{max} = 6.5$ ft (2.0 m), which is less than 12 ft (3.7 m), so a site-specific PTHA need not be performed (ASCE/SEI 6.5.2).

Step 3—Perform the R/H_T analysis   Sec. 3.8.2

An R/H_T analysis is not required because the runup elevation is obtained from the ASCE Geodatabase for the building in this example.

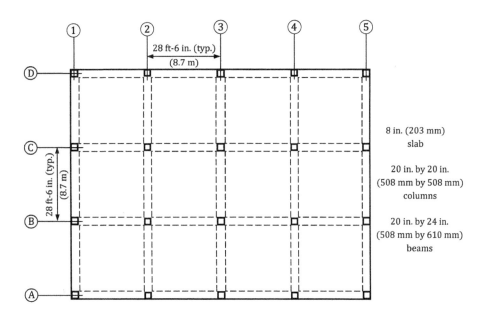

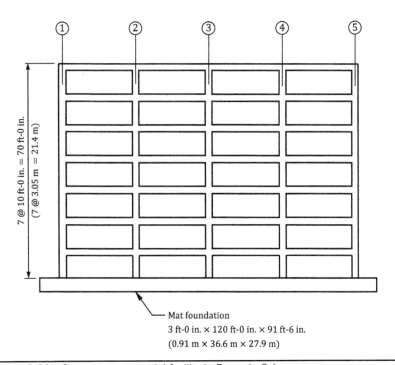

Figure 3.20   Seven-story essential facility in Example 3.1.

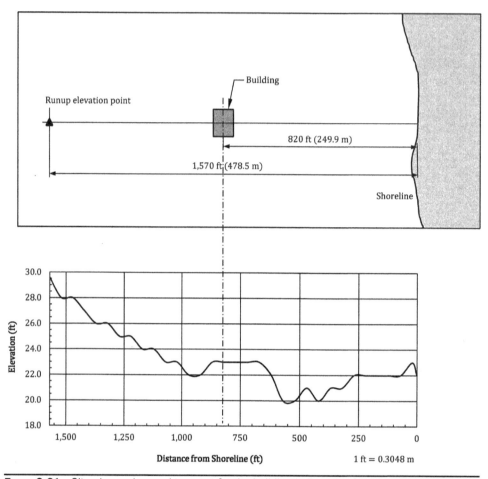

Figure 3.21 Site plan and ground transect for the building in Example 3.1.

Step 4—Perform the EGLA Sec. 3.8.3

The flowchart in Fig. 3.6 is used to determine the inundation depths and flow velocities along the ground transect in Fig. 3.21 and the information provided in the design data.

The analysis begins at the inundation point where $x_R = 1,570$ ft (478.5 m) and continues stepwise in the direction of the shoreline. The interval chosen in this example is 50 ft (15.2 m), which is less than the maximum limit of 100 ft (30.5 m) in ASCE/SEI 6.6.2.

At the runup point, the inundation depth is equal to zero, as are the velocity and energy (hydraulic head). An inundation depth of 0.1 ft (0.03 m) is used at this point so the hydraulic friction slope, s, can be calculated in a subsequent step (otherwise there would be division by zero).

A summary of the calculations to determine the inundation depths and flow velocities along the transect is given in Table 3.9.

Sample calculations at the building location, which is point 15 in Table 3.9, are given below using the flowchart in Fig. 3.6.

Point	Distance from Shoreline, x_i (ft)	Transect Elevation, z_i (ft)	Topographic Slope, φ	Froude Number, F_n	Friction Slope, s_i	Energy Head, E_g (ft)	Inundation Depth, h_i (ft)	Flow Velocity, u_i (ft/s)
Runup	1,570.0	29.6	0.0000	0.000	0.000000	0.00	0.10	0.00
1	1,520.0	28.0	0.0320	0.178	0.000896	1.64	1.62	1.29
2	1,470.0	28.0	0.0000	0.252	0.000708	1.68	1.63	1.83
3	1,420.0	27.0	0.0200	0.309	0.001060	2.73	2.61	2.83
4	1,370.0	26.0	0.0200	0.357	0.001208	3.79	3.57	3.82
5	1,320.0	26.0	0.0000	0.399	0.001360	3.86	3.58	4.28
6	1,270.0	25.0	0.0200	0.437	0.001631	4.94	4.51	5.27
7	1,220.0	25.0	0.0000	0.472	0.001761	5.03	4.53	5.70
8	1,170.0	24.0	0.0200	0.505	0.002010	6.13	5.44	6.68
9	1,120.0	24.0	0.0000	0.535	0.002127	6.24	5.46	7.10
10	1,070.0	23.0	0.0200	0.564	0.002361	7.36	6.35	8.07
11	1,020.0	23.0	0.0000	0.592	0.002470	7.48	6.36	8.47
12	970.0	22.0	0.0200	0.618	0.002692	8.61	7.23	9.43
13	920.0	22.0	0.0000	0.643	0.002795	8.75	7.25	9.83
14	870.0	23.0	−0.0200	0.668	0.003007	7.90	6.46	9.63
15	820.0	23.0	0.0000	0.691	0.003348	8.07	6.51	10.01
16	770.0	23.0	0.0000	0.714	0.003561	8.25	6.57	10.39
17	720.0	23.0	0.0000	0.736	0.003772	8.44	6.64	10.76
18	670.0	23.0	0.0000	0.757	0.003981	8.64	6.71	11.13
19	620.0	22.0	0.0200	0.778	0.004187	9.85	7.56	12.14
20	570.0	20.0	0.0400	0.798	0.004236	12.06	9.15	13.70
21	520.0	20.0	0.0000	0.818	0.004175	12.27	9.19	14.07
22	470.0	21.0	−0.0200	0.837	0.004366	11.49	8.51	13.85
23	420.0	20.0	0.0200	0.856	0.004684	12.72	9.31	14.82
24	370.0	21.0	−0.0200	0.874	0.004743	11.96	8.65	14.59
25	320.0	21.0	0.0000	0.892	0.005063	12.21	8.73	14.96
26	270.0	22.0	−0.0200	0.910	0.005249	11.47	8.11	14.71
27	220.0	22.0	0.0000	0.927	0.005586	11.75	8.22	15.08
28	170.0	22.0	0.0000	0.944	0.005768	12.04	8.33	15.46
29	120.0	22.0	0.0000	0.961	0.005948	12.34	8.44	15.84
30	70.0	22.0	0.0000	0.977	0.006125	12.64	8.56	16.22
31	20.0	23.0	−0.0200	0.994	0.006301	11.96	8.01	15.95
32	0.0	22.0	0.0500	1.000	0.006525	13.09	8.73	16.76

1 ft = 0.03048 m; 1 ft/s = 0.31 m/s

TABLE 3.9 EGLA Calculations for the Building in Example 3.1

Step 4a—Acquire x_R and R

$R = 29.6$ ft (9.0 m) is obtained from the ASCE Geodatabase at this site.

R occurs at $x_R = 1,570$ ft (478.5 m) from the shoreline.

Step 4b—Approximate the transect by a series of segmented slopes with transect points spaced not more than 100 ft (30.5 m).

The ground transect at this location is given in Fig. 3.21 with a maximum transect point spacing of 50 ft (15.2 m).

Step 4c—Calculate the slope within the segment

$$\varphi_{15} = \frac{z_{14} - z_{15}}{x_{14} - x_{15}} = \frac{23.0 - 23.0}{870.0 - 820.0} = 0.0000$$

Step 4d—Obtain Manning's coefficient, n, from ASCE/SEI Table 6.6-1 for the segment

Assume the frictional surface for this segment (as well as all the other segments in the transect) does not fall under any specific description in ASCE/SEI Table 6.6-1. Therefore, $n = 0.03$.

Step 4e—Calculate the Froude number, F_{r15}

$$F_{r15} = \alpha\left(1 - \frac{x_{15}}{x_R}\right)^{0.5} = 1.0 \times \left(1 - \frac{820.0}{1,570.0}\right)^{0.5} = 0.6912$$

Step 4f—Calculate the hydraulic friction slope, s_{15}

$$s_{15} = \frac{gF_{r15}^2}{\left(\dfrac{1.49}{n}\right)^2 h_{14}^{1/3}} = \frac{32.2 \times 0.6912^2}{\left(\dfrac{1.49}{0.03}\right)^2 \times (6.463)^{1/3}} = 0.003348$$

In S.I.:

$$s_{15} = \frac{gF_{r15}^2}{\left(\dfrac{1.00}{n}\right)^2 h_{14}^{1/3}} = \frac{9.81 \times 0.6912^2}{\left(\dfrac{1.00}{0.03}\right)^2 \times (1.97)^{1/3}} = 0.003365$$

Step 4g—Calculate the hydraulic head, $E_{g,15}$

$$E_{g,15} = E_{g,14} + (\varphi_{15} + s_{15})(x_{14} - x_{15})$$

$$= 7.90 + [(0.0000 + 0.003348) \times 50] = 8.07 \text{ ft}$$

In S.I.:

$$E_{g,15} = 2.41 + [(0.0000 + 0.003365) \times 15.2] = 2.46 \text{ m}$$

Step 4h—Calculate the inundation depth, h_{15}

$$h_{15} = \frac{E_{g,15}}{1+0.5F_{r15}^2} = \frac{8.07}{1+(0.5\times0.6912^2)} = 6.51 \text{ ft}$$

In S.I.:

$$h_{15} = \frac{E_{g,15}}{1+0.5F_{r15}^2} = \frac{2.46}{1+(0.5\times0.6912^2)} = 1.99 \text{ m}$$

Step 4i—Calculate the flow velocity, u_{15}

$$u_{15} = F_{r,15}(gh_{15})^{0.5} = 0.6912\times(32.2\times6.51)^{0.5} = 10.01 \text{ ft/s}$$

$$> 10.00 \text{ ft/s}$$

$$< \text{lesser of} \begin{cases} 1.5(gh_{15})^{0.5} = 21.72 \text{ ft/s} \\ 50.00 \text{ ft/s} \end{cases}$$

In S.I.:

$$u_{15} = F_{r,15}(gh_{15})^{0.5} = 0.6912\times(9.81\times1.99)^{0.5} = 3.05 \text{ m/s}$$

$$\geq 3.05 \text{ ft/s}$$

$$< \text{lesser of} \begin{cases} 1.5(gh_{15})^{0.5} = 6.63 \text{ m/s} \\ 15.24 \text{ m/s} \end{cases}$$

Plots of elevation, inundation depth, and flow velocity along the transect are given in Fig. 3.22.

The effects of sea level change in accordance with ASCE/SEI 6.5.3 are not considered in this example, but should be included wherever appropriate. Additionally, it is assumed the flow velocity amplification requirements in ASCE/SEI 6.8.5 are not applicable.

Similar analyses must be performed for transects located 22.5 degrees on either side of the transect in this example (see ASCE/SEI 6.8.6.1 and Fig. 3.7).

Step 5—Determine the hydrostatic loads   Sec. 3.9

Step 5a—Buoyancy loads

Buoyancy loads are considered in load case 1 where the inundation depth is equal to the following (ASCE/SEI 6.8.3.1):

$$h_{sx} = \text{lesser of} \begin{cases} h_{max} = 6.5 \text{ ft (2.0 m)} \\ \text{Story height} = 10.0 \text{ ft (3.1 m)} \\ \text{Height to top of first story window} = 10.0-2.0 = 8.0 \text{ ft (2.4 m)} \end{cases}$$

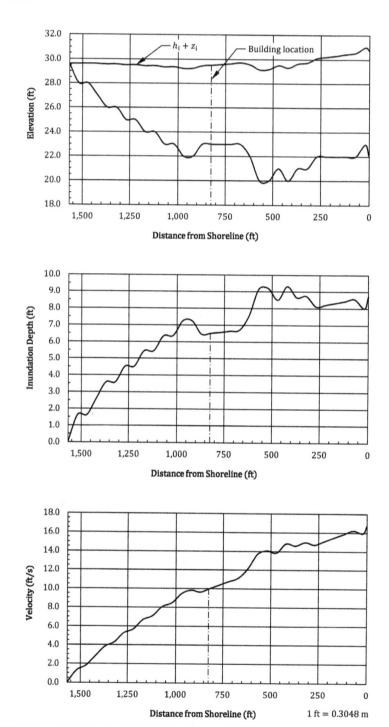

FIGURE 3.22   Elevation, inundation depth, and flow velocity along the transect in Example 3.1.

The volume of displaced water, V_w, in this case consists of the volume of the inundated building plus the volume of the mat foundation, which is connected to the building columns.

$$F_v = \gamma_s V_w$$

$$= (1.1 \times 64.0) \times [(6.5 \times 115.7 \times 87.2) + (3.0 \times 120.0 \times 91.5)]/1,000 \qquad (3.4)$$

$$= 6,936 \text{ kips}$$

In S.I.:

$$F_v = \gamma_s V_w = (1.1 \times 10.05) \times [(2.0 \times 35.3 \times 26.6) + (0.91 \times 36.6 \times 27.9)] = 31,034 \text{ kN}$$

Step 5b—Unbalanced lateral hydrostatic loads

This load is not applicable in this example because there are no inundated exterior walls.

Step 5c—Residual water surcharge load on floors and walls

Because $h_{max} = 6.5$ ft (2.0 m) is less than the first story height, there are no residual water surcharge loads on the floors.

Step 5d—Hydrostatic surcharge pressure on foundation

$$p_s = \gamma_s h_{max} = (1.1 \times 64.0) \times 6.5 = 458 \text{ lb/ft}^2 \qquad (3.7)$$

In S.I.:

$$p_s = \gamma_s h_{max} = (1.1 \times 10.05) \times 2.0 = 22.1 \text{ kN/m}^2$$

Step 6—Determine the hydrodynamic loads Sec. 3.10

Step 6a—Option 1: Simplified uniform lateral static pressure

$$p_{uw} = 1.25 I_{tsu} \gamma_s h_{max} = 1.25 \times 1.25 \times (1.1 \times 64.0) \times 6.5 = 715 \text{ lb/ft}^2 \qquad (3.8)$$

This pressure is applied over the width of the building and over a height of $1.3 h_{max} = 8.5$ ft

In S.I.:

$$p_{uw} = 1.25 I_{tsu} \gamma_s h_{max} = 1.25 \times 1.25 \times (1.1 \times 10.05) \times 2.0 = 34.6 \text{ kN/m}^2$$

This pressure is applied over the width of the building and over a height of $1.3 h_{max} = 2.6$ m.

$$\text{Total force} = 715 \times 115.7 \times 8.5/1,000 = 703 \text{ kips}$$

In S.I.:

$$\text{Total force} = 34.6 \times 35.3 \times 2.6 = 3,176 \text{ kN}$$

The simplified uniform lateral static pressure is depicted in Fig. 3.23.

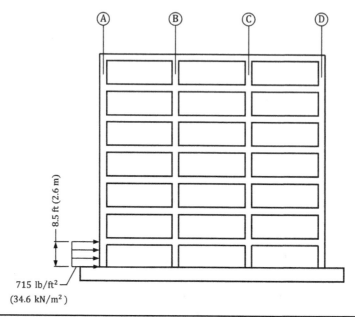

FIGURE 3.23 Simplified uniform lateral static hydrodynamic pressure in Example 3.1.

Step 6b—Option 2: Detailed hydrodynamic lateral forces   Table 3.5
  • Overall drag force on building
    The flowchart in Fig. 3.13 is used to determine F_{dx} for load cases 1, 2, and 3.

$$\rho_s = 2.2 \text{ slugs/ft}^3 \ (1{,}128 \text{ kg/m}^3)$$

For Tsunami Category IV buildings, $I_{tsu} = 1.25$ from ASCE/SEI Table 6.8-1. Values of C_d for load cases 1, 2, and 3 are given in Table 3.10. Sample calculations for load case 1 are as follows:

$$h_{sx} = \text{lesser of} \begin{cases} h_{max} = 6.5 \text{ ft (2.0 m)} \\ \text{Story height} = 10.0 \text{ ft (3.1 m)} \\ \text{Height to top of first story window} = 10.0 - 2.0 = 8.0 \text{ ft (2.4 m)} \end{cases}$$

$B/h_{sx} = 115.7/6.5 = 17.8$

Load Case	h_{sx}, ft (m)	B/h_{sx}	C_d
1	6.5 (2.0)	17.8	1.32
2	4.3 (1.3)	26.9	1.41
3	6.5 (2.0)	17.8	1.32

TABLE 3.10 Values of C_d for Load Cases 1, 2, and 3 in Example 3.1

Load Case	h_{sx}, ft (m)	C_{cx}
1	6.5 (2.0)	1.0
2	4.3 (1.3)	0.7
3	6.5 (2.0)	0.7

TABLE 3.11   Values of C_{cx} for Load Cases 1, 2, and 3 in Example 3.1

From ASCE/SEI Table 6.10-1, $C_d = 1.32$ by linear interpolation
    Values of C_{cx} for load cases 1, 2, and 3 are given in Table 3.11. Sample calculations for load cases 1 and 2 are as follows:

For load case 1:

$$C_{cx} = 1.0 \text{ because it is assumed the exterior walls have not failed}$$

For load case 2:

$$h_{sx} = 2h_{max}/3 = 4.3 \text{ ft (1.3 m)}$$

$$A_{col} = (20.0/12) \times 4.3 = 7.2 \text{ ft}^2 \ (0.67 \text{ m}^2)$$

$A_{wall} = 0$ (there are no walls)
$A_{beam} = 0$ (the beams are above the inundation depth)

$$C_{cx} = \frac{\sum(A_{col} + A_{wall}) + 1.5A_{beam}}{Bh_{sx}} = \frac{20 \times 7.2}{115.7 \times 4.3} = 0.3 < 0.7, \text{ use } 0.7$$

In S.I.:

$$C_{cx} = \frac{20 \times 0.67}{35.3 \times 1.3} = 0.3 < 0.7, \text{ use } 0.7$$

The overall drag force, F_{dx}, for load case 1 is as follows where the minimum flow velocity of 10.0 ft/s (3.1 m/s) is used:

$$F_{dx} = \frac{1}{2}\rho_s I_{tsu} C_d C_{cx} B(hu^2)$$

$$= \frac{1}{2} \times 2.2 \times 1.25 \times 1.32 \times 1.0 \times 115.7 \times (6.5 \times 10.0^2)/1,000 = 137 \text{ kips}$$

In S.I.:

$$F_{dx} = \frac{1}{2} \times (1,128/1,000) \times 1.25 \times 1.32 \times 1.0 \times 35.3 \times (2.0 \times 3.1^2) = 631 \text{ kN}$$

The hydrodynamic and buoyancy loads in load case 1 are shown in Fig. 3.24.

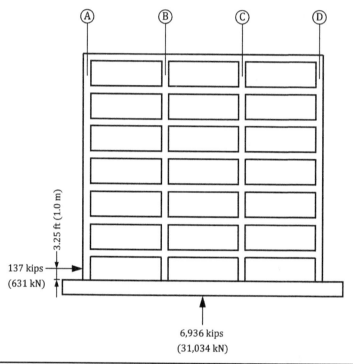

137 kips
(631 kN)

3.25 ft (1.0 m)

6,936 kips
(31,034 kN)

FIGURE 3.24   Load case 1 for the building in Example 3.1.

For load case 2:

$$F_{dx} = \frac{1}{2}\rho_s I_{tsu}C_dC_{cx}B(hu^2)$$

$$= \frac{1}{2}\times 2.2\times 1.25\times 1.41\times 0.7\times 115.7\times(4.3\times 10.0^2)/1,000 = 68 \text{ kips}$$

In S.I.:

$$F_{dx} = \frac{1}{2}\times(1,128/1,000)\times 1.25\times 1.41\times 0.7\times 35.3\times(1.3\times 3.1^2) = 307 \text{ kN}$$

This force is applied to the building at $(4.33/2) = 3.17$ ft (0.66 m) above the ground level, as shown in Fig. 3.25.

For load case 3, $u = u_{max}/3 = 10.0/3 = 3.3$ ft/s (1.0 m/s) < 10.0 ft/s (3.1 m/s) and $h = h_{max} = 6.5$ ft (2.0 m). Therefore, with $u = 10.0$ ft/s (3.1 m/s), $C_d = 1.32$, and $C_{cx} = 0.7$, $F_{dx} = 96$ kips (442 kN). This force is applied at $(6.5/2) = 3.25$ ft (1.0 m) above the base of the column (see Fig. 3.26).

- Drag force on components

The flowchart in Fig. 3.15 is used to determine F_d for the interior and exterior columns.

$$\rho_s = 2.2 \text{ slugs/ft}^3 \text{ (1,128 kg/m}^3\text{)}$$

For tsunami category IV buildings, $I_{tsu} = 1.25$ from ASCE/SEI Table 6.8-1.

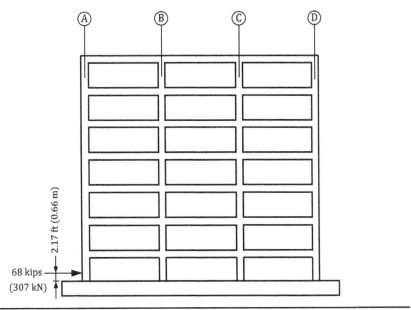

68 kips
(307 kN)

2.17 ft (0.66 m)

Figure 3.25 Load case 2 for the building in Example 3.1.

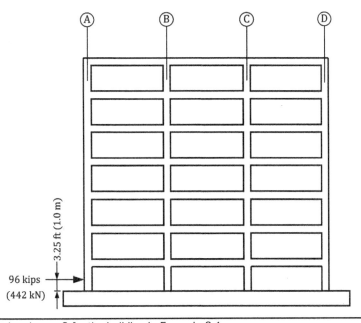

96 kips
(442 kN)

3.25 ft (1.0 m)

Figure 3.26 Load case 3 for the building in Example 3.1.

Interior columns:

For square columns, $C_d = 2.0$ from ASCE/SEI Table 6.10-2.

$$b = 20.0/12 = 1.7 \text{ ft (0.51 m)}$$

Load case 2 governs where inundated height $h_e = 6.5$ ft (2.0 m) and $u = u_{max} = 10.0$ ft/s (3.1m/s):

$$F_d = \frac{1}{2} \rho_s I_{tsu} C_d b (h_e u^2) \tag{3.10}$$

$$= \frac{1}{2} \times 2.2 \times 1.25 \times 2.0 \times 1.7 \times (6.5 \times 10.0^2)/1,000 = 3 \text{ kips}$$

This load is applied as an equivalent uniformly distributed lateral load of $3/6.5 = 0.5$ kips/ft over the inundated portion of the column (see Fig. 3.27).

In S.I.:

$$F_d = \frac{1}{2} \times (1,128/1,000) \times 1.25 \times 2.0 \times 0.51 \times (2.0 \times 3.1^2) = 13.8 \text{ kN}$$

This load is applied as an equivalent uniformly distributed lateral load of $13.8/2.0 = 6.9$ kN/m over the inundated portion of the column.

Exterior columns:

$$C_d = 2.0 \qquad\qquad \text{ASCE/SEI 6.10.2.2}$$

Tributary width of an exterior column = 28.5 ft (8.7 m)

$$C_{cx} = 0.7$$

$$b = C_{cx} \times \text{Tributary width} = 0.7 \times 28.5 = 20.0 \text{ ft (6.1 m)}$$

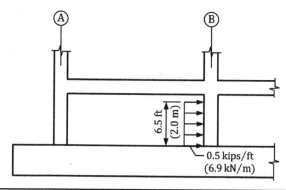

FIGURE 3.27 Drag force on an interior column in the building in Example 3.1.

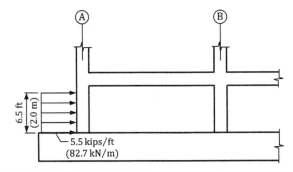

FIGURE 3.28   Drag force on an exterior column in the building in Example 3.1.

Load case 2 governs where inundated height $h_e = 6.5$ ft (2.0 m) and $u = u_{max} = 10.0$ ft/s (3.1 m/s):

$$F_d = \frac{1}{2} \rho_s I_{tsu} C_d b(h_e u^2) \qquad \text{Eq. (3.10)}$$

$$= \frac{1}{2} \times 2.2 \times 1.25 \times 2.0 \times 20.0 \times (6.5 \times 10.0^2)/1,000 = 36 \text{ kips}$$

This load is applied as an equivalent uniformly distributed lateral load of $36/6.5 = 5.5$ kips/ft over the inundated portion of the column (see Fig. 3.28).

In S.I.:

$$F_d = \frac{1}{2} \times (1,128/1,000) \times 1.25 \times 2.0 \times 6.1 \times (2.0 \times 3.1^2) = 165.3 \text{ kN}$$

This load is applied as an equivalent uniformly distributed lateral load of $165.3/2.0 = 82.7$ kN/m over the inundated portion of the column.

- Tsunami loads on vertical structural components

  This load is not applicable in this example because there are no vertical structural components.

- Hydrodynamic loads on perforated walls

  This load is not applicable in this example because there are no perforated walls.

- Walls angled to the flow

  This load is not applicable in this example because there are no walls angled to the flow.

- Hydrodynamic pressure associated with slabs

  The loads associated with (1) spaces in buildings subjected to flow stagnation pressurization, (2) hydrodynamic surge uplift at horizontal slabs, and (3) tsunami pore flow entrapped in structural wall-slab recess are not applicable in this example.

Step 7—Determine the debris impact loads Sec. 3.11

Because the inundation depth exceeds 3 ft (0.91 m) at the site, the exterior columns below the flow depth are subjected to debris impact loads in accordance with ASCE/SEI 6.11.

Step 7a—Option 1: Alternative simplified debris impact static load

$$F_i = 330C_oI_{tsu} = 330 \times 0.65 \times 1.25 = 268 \text{ kips} \tag{3.15a}$$

In S.I.:

$$F_i = 1,470C_oI_{tsu} = 1,470 \times 0.65 \times 1.25 = 1,194 \text{ kN} \tag{3.15b}$$

Because the site is located in an impact zone for shipping containers, it is not permitted to reduce the calculated value of the impact force by 50 percent.

The impact force is applied to an exterior column at points along the height of the column critical for flexure and shear. Illustrated in Fig. 3.29 is the impact force applied at 4 ft (1.2 m) from the base of the column, which is the critical section for flexure.

Step 7b—Option 2: Detailed debris impact loads

• Wood logs and poles

The flowchart in Fig. 3.17 is used to determine the impact force due to wood logs and poles.

$$u_{max} = 10.0 \text{ ft/s (3.1 m/s)}$$

Minimum log stiffness = 350 kip/in. (61,300 kN/m)

Assuming the column has simple connections at both ends, the stiffness of the column is determined as follows:

$$k = \frac{48EI}{L^3} = \frac{48 \times (3,605,000/1,000) \times \left(\frac{1}{12} \times 20^4\right)}{(8.0 \times 12)^3} = 2,608 \text{ kip/in.} > 350 \text{ kip/in.}$$

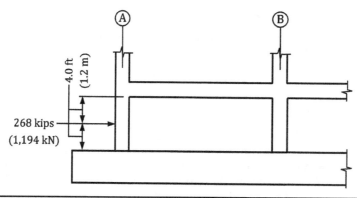

FIGURE 3.29 Alternative simplified debris impact static load on an exterior column in the building in Example 3.1.

In S.I.:

$$k = \frac{48 \times (24{,}900 \times 10^3) \times \left[\frac{1}{12} \times \left(\frac{508}{1{,}000}\right)^4\right]}{(2.44)^3} = 456{,}609 \text{ kN/m} > 61{,}300 \text{ kN/m}$$

Therefore, use $k = 350$ kip/in. (61,300 kN/m).
Assume minimum debris weight $W_d = 1{,}000$ lb (4.5 kN).
Debris mass, m_d:

$$m_d = W_d/g = 1{,}000/32.2 = 31.1 \text{ slugs}$$

In S.I.:

$$m_d = 4.5 \times 10^3/9.81 = 459 \text{ kg}$$

Nominal debris impact force, F_{ni}:

$$F_{ni} = u_{max}\sqrt{km_d} = 10.0 \times \sqrt{(350 \times 12) \times (31.1/1{,}000)} = 114 \text{ kips}$$

In S.I.:

$$F_{ni} = 3.1 \times \sqrt{(61{,}300/1{,}000) \times 459} = 520 \text{ kN}$$

For Tsunami Category IV buildings, $I_{tsu} = 1.25$ from ASCE/SEI Table 6.8-1.
Orientation coefficient $C_o = 0.65$.
Design debris impact force $F_i = I_{tsu}C_oF_{ni} = 1.25 \times 0.65 \times 114 = 93$ kips.
In S.I.:

$$\text{Design debris impact force } F_i = 1.25 \times 0.65 \times 520 = 423 \text{ kN}$$

Impulse duration for elastic impact, t_d:

$$t_d = \frac{2m_d u_{max}}{F_{ni}} = \frac{2 \times 31.1 \times 10.0}{114 \times 1{,}000} = 0.006 \text{ s}$$

In S.I.:

$$t_d = \frac{2m_d u_{max}}{F_{ni}} = \frac{2 \times 459 \times 3.1}{520 \times 1{,}000} = 0.006 \text{ s}$$

Assuming the column has simple connections at both ends, the natural period of the column is determined as follows:

$$\text{Mass of column per unit length } m = \frac{20 \times 20}{144} \times \frac{150}{32.2} = 12.9 \text{ slugs/ft}$$

$$T_n = \frac{2L^2}{\pi\sqrt{\frac{EI}{m}}} = \frac{2 \times 8.0^2}{\pi\sqrt{\frac{(3{,}605{,}000 \times 144) \times \left[\frac{1}{12} \times \left(\frac{20}{12}\right)^4\right]}{12.9}}} = 0.008 \text{ s}$$

In S.I.:

$$m = \left(\frac{508}{1,000}\right)^2 \times \frac{(23.6 \times 10^3)}{9.81} = 621 \text{ kg/m}$$

$$T_n = \frac{2 \times 2.4^2}{\pi \sqrt{\dfrac{(29,600 \times 10^6) \times \left[\dfrac{1}{12} \times \left(\dfrac{508}{1,000}\right)^4\right]}{621}}} = 0.007 \text{ s}$$

$$t_d/T_n = 0.006/0.008 = 0.75$$

From ASCE/SEI Table 6.11-1, $R_{max} = 1.8$
Dynamic impact force $= R_{max}F_i = 1.8 \times 93 = 167$ kips

In S.I.:

$$t_d/T_n = 0.006/0.007 = 0.86$$

From ASCE/SEI Table 6.11-1, $R_{max} = 1.8$

$$\text{Dynamic impact force} = R_{max}F_i = 1.8 \times 423 = 761 \text{ kN}$$

The dynamic impact force is applied anywhere from 3 ft (0.91 m) up to $h_{max} = 6.5$ ft (2.0 m) from the base of the column. Illustrated in Fig. 3.30 is the dynamic impact force applied at 4 ft (1.2 m) from the base of the column, which is the critical section for flexure.

- Vehicles

$$F_i = 30I_{tsu} = 30 \times 1.25 = 38 \text{ kips}$$

In S.I.:

$$F_i = 130I_{tsu} = 130 \times 1.25 = 163 \text{ kN}$$

This impact force is applied anywhere from 3 ft (0.91 m) up to $h_{max} = 6.5$ ft (2.0 m) from the base of the column.

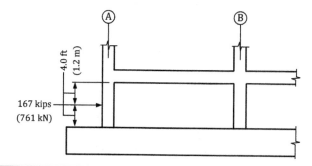

FIGURE 3.30   Dynamic impact force due to wood logs and poles in Example 3.1.

- Tumbling boulders and concrete debris

  This impact load must be considered because $h_{max} = 6.5$ ft (2.0 m) > 6.0 ft (1.8 m).

  $$F_i = 8,000I_{tsu} = 8,000 \times 1.25/1,000 = 10 \text{ kips}$$

  In S.I.:

  $$F_i = 36I_{tsu} = 36 \times 1.25 = 45 \text{ kN}$$

  This impact force is applied 2 ft (0.61 m) above the base of the column.

- Shipping containers

  The flowchart in Fig. 3.18 is used to determine the impact load due to shipping containers.

  $$u_{max} = 10.0 \text{ ft/s} (3.1 \text{ m/s})$$

  For a 20-ft (6.1-m) shipping container, stiffness = 245 kip/in. (42,900 kN/m).

  Assuming the column has simple connections at both ends, the stiffness of the column is determined as follows:

  $$k = \frac{48EI}{L^3} = \frac{48 \times (3,605,000/1,000) \times \left(\frac{1}{12} \times 20^4\right)}{(8.0 \times 12)^3}$$

  $$= 2,608 \text{ kip/in.} > 245 \text{ kip/in.}$$

  In S.I.:

  $$k = \frac{48 \times (24,900 \times 10^3) \times \left[\frac{1}{12} \times \left(\frac{508}{1,000}\right)^4\right]}{(2.44)^3} = 456,609 \text{ kN/m} > 42,900 \text{ kN/m}$$

  Therefore, use $k = 245$ kip/in. (42,900 kN/m).

  $$\text{Empty debris weight } W_d = 5,000 \text{ lb (22.2 kN)}$$

  Loaded debris weight $W_d = 29,000$ lb (129.0 kN)
  Debris mass of the empty shipping container:

  $$m_d = W_d/g = 5,000/32.2 = 155.3 \text{ slugs}$$

  In S.I.:

  $$m_d = W_d/g = 22.2 \times 1,000/9.81 = 2,263 \text{ kg}$$

  Nominal debris impact force, F_{ni}:

  $$F_{ni} = u_{max}\sqrt{km_d} = 10.0 \times \sqrt{(245 \times 12) \times (155.3/1,000)} = 214 \text{ kips} < 220 \text{ kips}$$

In S.I.:

$$F_{ni} = 3.1 \times \sqrt{(42,900/1,000) \times 2,263} = 966 \text{ kN} < 980 \text{ kN}$$

For Tsunami Category IV buildings, $I_{tsu} = 1.25$ from ASCE/SEI Table 6.8-1.

Orientation coefficient $C_o = 0.65$.

Design debris impact force $F_i = I_{tsu} C_o F_{ni} = 1.25 \times 0.65 \times 214 = 174$ kips

In S.I.:

Design debris impact force $F_i = 1.25 \times 0.65 \times 966 = 785$ kN

Impact force for an empty shipping container:

Impulse duration for elastic impact $t_d = \dfrac{2m_d u_{max}}{F_{ni}} = \dfrac{2 \times 155.3 \times 10.0}{214 \times 1,000} = 0.015$ s

Assuming the column has simple connections at both ends, the natural period of the column is determined above for wood logs and poles as 0.008 s (the natural period is equal to 0.007 s when S.I. units are used; see the calculations under the wood logs and poles section of this example).

$$t_d/T_n = 0.015/0.008 = 1.88$$

From ASCE/SEI Table 6.11-1, $R_{max} = 1.5$

Dynamic impact force $= R_{max} F_i = 1.5 \times 174 = 261$ kips

In S.I.:

$$t_d/T_n = 0.015/0.007 = 2.14$$

From ASCE/SEI Table 6.11-1, $R_{max} = 1.5$

Dynamic impact force $= R_{max} F_i = 1.5 \times 785 = 1,178$ kN

Impact force for a loaded shipping container:

Impulse duration for elastic impact, t_d:

$$t_d = \frac{(m_d + m_{contents})u_{max}}{F_{ni}} = \frac{(29,000/32.2) \times 10.0}{214 \times 1,000} = 0.042 \text{ s}$$

$$t_d/T_n = 0.042/0.008 = 5.25$$

From ASCE/SEI Table 6.11-1, $R_{max} = 1.5$

Dynamic impact force $= R_{max} F_i = 1.5 \times 174 = 261$ kips

In S.I.:

$$t_d/T_n = 0.042/0.007 = 6.00$$

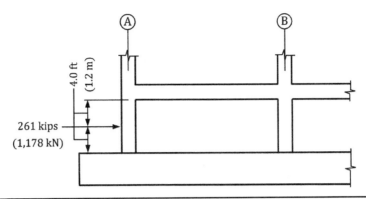

FIGURE 3.31   Dynamic impact force due to shipping containers in Example 3.1.

From ASCE/SEI Table 6.11-1, $R_{max} = 1.5$

$$\text{Dynamic impact force} = R_{max}F_i = 1.5 \times 785 = 1,178 \text{ kN}$$

In this example, the dynamic impact forces of the empty and fully load shipping containers are equal.

The dynamic impact force is applied anywhere from 3 ft (0.91 m) up to $h_{max} = 6.5$ ft (2.0 m) from the base of the column. Illustrated in Fig. 3.31 is the dynamic impact force applied at 4 ft (1.2 m) from the base of the column, which is the critical section for flexure.

Step 8—Determine the tsunami load combinations   Sec. 3.7.2

Principal tsunami forces and effects on the lateral force–resisting system and the individual components must be combined with other specified loads using the load combinations in ASCE/SEI 6.8.3.3 [see Eqs. (3.1) and (3.2) of this publication]. As noted previously, the debris impact loads need not be combined with other tsunami-related loads determined in accordance with ASCE/SEI Chapter 6.

3.13.2   Example 3.2—Six-Story Residential Building

Determine the tsunami loads, F_{TSU}, on the six-story residential in Fig. 3.32. Assume the following:

- Building site:
  - Latitude: 19.71717°
  - Longitude: −155.06650°
- The building is located 4,287 ft (1,307 m) from the shoreline as shown in Fig. 3.33.
- The ground transect perpendicular to the shoreline is given in Fig. 3.33.
- Runup elevation is not given in the ASCE Geodatabase.
- The mean slope angle of the nearshore profile, Φ, is equal to 15 degrees.

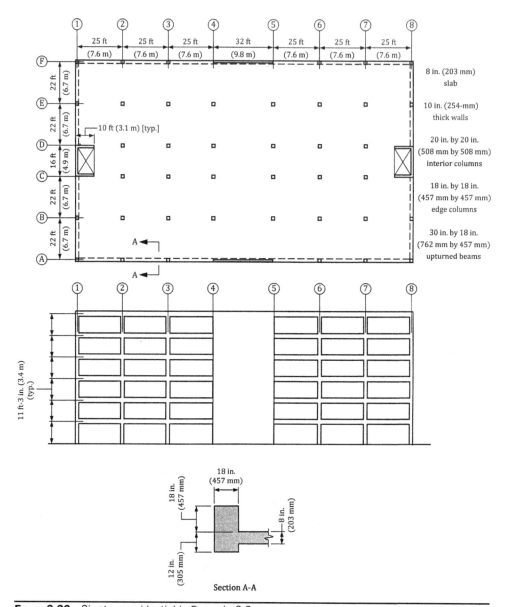

FIGURE 3.32 Six-story residential in Example 3.2.

- Offshore tsunami amplitude, H_T, from the ASCE Geodatabase is equal to 28 ft (8.5 m).
- Predominant wave period, T_{TSU}, from the ASCE Geodatabase is equal to 16 minutes.
- Tsunami bores are expected within this area.
- The site is not located in an impact zone for shipping containers in accordance with ASCE/SEI 6.11.5.

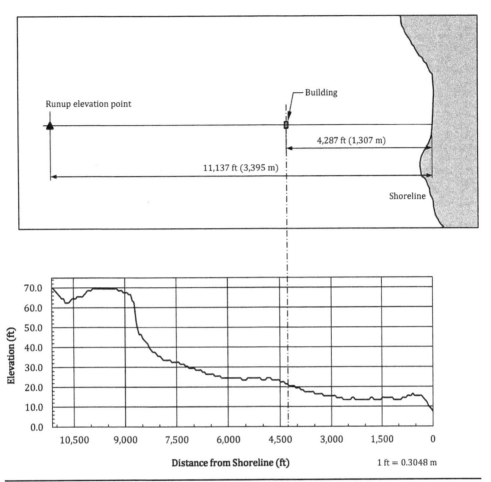

Figure 3.33 Site plan and ground transect for the building in Example 3.2.

- Windows are not designed for large impact loads or blast loads and span from the top of the slab to the underside of the beams.
- The building is supported by pile foundations beneath the columns and walls.
- The slab-on-grade is isolated from the columns and walls.
- The perimeter beams are upturned as shown in Detail A-A in Fig. 3.32.
- The structural members are constructed of reinforced concrete with a unit weight equal to 150 lb/ft³ (23.6 kN/m³) and a modulus of elasticity, E, equal to 3,605,000 lb/in.² (24,900 MPa).

Solution

Step 1—Determine the TRC Table 3.2

The building is a residential building that has been designated by the local authorities to be designed for tsunami effects, so the TRC is II.

Step 2—Determine the permitted analysis procedure Sec. 3.8

As noted in the design data, the runup elevation is not given in the ASCE Geodatabase, so the permitted analysis procedures are given in Table 3.4.

For a TRC II building, the R/H_T and EGLA methods are both required. It is determined in Step 4 below from an EGLA that $h_{max} = 44.5$ ft (13.6 m), which is greater than 3 ft (0.91 m), so the requirements in ASCE/SEI Chapter 6 are applicable.

Step 3—Perform the R/H_T analysis Sec. 3.8.2

$$\cot \Phi = \cot 15° = 3.73$$

Determine the surf similarity parameter:

$$\xi_{100} = \frac{T_{TSU}}{\cot \Phi}\sqrt{\frac{g}{2\pi H_T}} = \frac{16 \times 60}{3.73} \times \sqrt{\frac{32.2}{2\pi \times 28.0}} = 110.1 \qquad (3.3)$$

In S.I.:

$$\xi_{100} = \frac{T_{TSU}}{\cot \Phi}\sqrt{\frac{g}{2\pi H_T}} = \frac{16 \times 60}{3.73} \times \sqrt{\frac{9.81}{2\pi \times 8.53}} = 110.1$$

ASCE/SEI Equation (6.5-2e) is used to determine R/H_T because $\xi_{100} > 100$:

$$R/H_T = 2.5$$

Therefore, $R = 2.5 \times 28.0 = 70.0$ ft (21.3 m).

From the transect, this occurs at 11,137 ft (3,395 m) from the shoreline (see Fig. 3.33).

Step 4—Perform the EGLA Sec. 3.8.3

The flowchart in Fig. 3.6 is used to determine the inundation depths and flow velocities along the ground transect in Fig. 3.33 and the information provided in the design data.

The analysis begins at the inundation point where $x_R = 11,137$ ft (3,395 m) and continues stepwise in the direction of the shoreline. The interval chosen in this example is 50 ft (15.2 m), which is less than the maximum limit of 100 ft (30.5 m) in ASCE/SEI 6.6.2.

At the runup point, the inundation depth is equal to zero, as are the velocity and energy (hydraulic head). An inundation depth of 0.1 ft (0.03 m) is used at this point, so the hydraulic friction slope, s, can be calculated in a subsequent step (otherwise, there would be division by zero).

A summary of the calculations to determine the inundation depths and flow velocities along the transect is given in Table 3.12.

Sample calculations at the building location, which is point 137 in Table 3.12, are given below using the flowchart in Fig. 3.6.

Point	Distance from Shoreline, x_i (ft)	Transect Elevation, z_i (ft)	Topographic Slope, φ	Froude Number, F_n	Friction Slope, s_i	Energy Head, E_g (ft)	Inundation Depth, h_i (ft)	Flow Velocity, u_i (ft/s)
Runup	11,137.0	70.0	0.0000	0.000	0.000000	0.00	0.10	0.00
1	11,087.0	69.5	0.0100	0.067	0.000126	0.51	0.51	0.27
2	11,037.0	68.5	0.0200	0.095	0.000147	1.51	1.51	0.66
3	10,987.0	67.5	0.0200	0.116	0.000153	2.52	2.50	1.04
4	10,937.0	66.5	0.0200	0.134	0.000173	3.53	3.50	1.42
5	10,887.0	65.5	0.0200	0.150	0.000193	4.54	4.49	1.80
6	10,837.0	64.5	0.0200	0.164	0.000213	5.55	5.48	2.18
7	10,787.0	64.5	0.0000	0.177	0.000233	5.56	5.48	2.35
8	10,737.0	62.5	0.0400	0.190	0.000266	7.58	7.44	2.93
9	10,687.0	62.5	0.0000	0.201	0.000270	7.59	7.44	3.11
10	10,637.0	62.5	0.0000	0.212	0.000300	7.60	7.44	3.28
11	10,587.0	63.5	−0.0200	0.222	0.000330	6.62	6.46	3.21
12	10,537.0	64.5	−0.0200	0.232	0.000378	5.64	5.49	3.09
13	10,487.0	64.5	0.0000	0.242	0.000432	5.66	5.50	3.22
14	10,437.0	64.5	0.0000	0.251	0.000465	5.68	5.51	3.34
15	10,387.0	65.5	−0.0200	0.260	0.000498	4.71	4.56	3.14
16	10,337.0	65.5	0.0000	0.268	0.000566	4.74	4.57	3.25
17	10,287.0	65.5	0.0000	0.276	0.000600	4.77	4.59	3.36
18	10,237.0	65.5	0.0000	0.284	0.000635	4.80	4.61	3.46
19	10,187.0	66.5	−0.0200	0.292	0.000669	3.83	3.68	3.18
20	10,137.0	67.5	−0.0200	0.300	0.000759	2.87	2.75	2.82
21	10,087.0	68.5	−0.0200	0.307	0.000879	1.91	1.83	2.36
22	10,037.0	68.5	0.0000	0.314	0.001054	1.97	1.87	2.44
23	9,987.0	68.5	0.0000	0.321	0.001093	2.02	1.92	2.53
24	9,937.0	69.5	−0.0200	0.328	0.001131	1.08	1.02	1.88
25	9,887.0	69.5	0.0000	0.335	0.001454	1.15	1.09	1.98
26	9,837.0	69.5	0.0000	0.342	0.001481	1.22	1.16	2.09
27	9,787.0	69.5	0.0000	0.348	0.001507	1.30	1.23	2.19
28	9,737.0	69.5	0.0000	0.355	0.001533	1.38	1.30	2.29
29	9,687.0	69.5	0.0000	0.361	0.001559	1.45	1.37	2.39
30	9,637.0	69.5	0.0000	0.367	0.001585	1.53	1.44	2.50
31	9,587.0	69.5	0.0000	0.373	0.001610	1.61	1.51	2.60
32	9,537.0	69.5	0.0000	0.379	0.001635	1.70	1.58	2.71
33	9,487.0	69.5	0.0000	0.385	0.001660	1.78	1.66	2.81

1 ft = 0.3048 m; 1 ft/s = 0.31 m/s

TABLE 3.12 EGLA Calculations for the Building in Example 3.2

Point	Distance from Shoreline, x_i (ft)	Transect Elevation, z_i (ft)	Topographic Slope, φ	Froude Number, F_n	Friction Slope, s_i	Energy Head, E_g (ft)	Inundation Depth, h_i (ft)	Flow Velocity, u_i (ft/s)
34	9,437.0	69.5	0.0000	0.391	0.001684	1.86	1.73	2.92
35	9,387.0	69.5	0.0000	0.396	0.001708	1.95	1.81	3.02
36	9,337.0	69.5	0.0000	0.402	0.001732	2.04	1.88	3.13
37	9,287.0	69.5	0.0000	0.408	0.001756	2.12	1.96	3.24
38	9,237.0	69.5	0.0000	0.413	0.001779	2.21	2.04	3.35
39	9,187.0	68.5	0.0200	0.418	0.001803	3.30	3.04	4.14
40	9,137.0	68.5	0.0000	0.424	0.001619	3.38	3.10	4.24
41	9,087.0	68.5	0.0000	0.429	0.001647	3.47	3.17	4.34
42	9,037.0	67.5	0.0200	0.434	0.001675	4.55	4.16	5.02
43	8,987.0	67.5	0.0000	0.439	0.001567	4.63	4.22	5.12
44	8,937.0	67.5	0.0000	0.444	0.001596	4.71	4.28	5.22
45	8,887.0	66.5	0.0200	0.449	0.001624	5.79	5.26	5.85
46	8,837.0	66.5	0.0000	0.454	0.001550	5.87	5.32	5.95
47	8,787.0	63.5	0.0600	0.459	0.001578	8.95	8.09	7.41
48	8,737.0	62.5	0.0200	0.464	0.001401	10.02	9.04	7.92
49	8,687.0	54.5	0.1600	0.469	0.001378	18.08	16.29	10.74
50	8,637.0	49.5	0.1000	0.474	0.001156	23.14	20.81	12.26
51	8,587.0	46.5	0.0600	0.479	0.001087	26.20	23.51	13.16
52	8,537.0	46.5	0.0000	0.483	0.001064	26.25	23.51	13.29
53	8,487.0	44.5	0.0400	0.488	0.001084	28.30	25.29	13.92
54	8,437.0	43.5	0.0200	0.492	0.001078	29.36	26.18	14.30
55	8,387.0	42.5	0.0200	0.497	0.001085	30.41	27.07	14.67
56	8,337.0	41.5	0.0200	0.501	0.001093	31.47	27.95	15.04
57	8,287.0	39.5	0.0400	0.506	0.001101	33.52	29.72	15.65
58	8,237.0	38.5	0.0200	0.510	0.001097	34.58	30.59	16.02
59	8,187.0	37.5	0.0200	0.515	0.001106	35.63	31.46	16.38
60	8,137.0	37.5	0.0000	0.519	0.001114	35.69	31.45	16.52
61	8,087.0	36.5	0.0200	0.523	0.001133	36.74	32.32	16.88
62	8,037.0	35.5	0.0200	0.528	0.001141	37.80	33.18	17.25
63	7,987.0	35.5	0.0000	0.532	0.001149	37.86	33.17	17.38
64	7,937.0	34.5	0.0200	0.536	0.001167	38.92	34.03	17.74
65	7,887.0	33.5	0.0200	0.540	0.001176	39.98	34.89	18.11
66	7,837.0	33.5	0.0000	0.544	0.001184	40.03	34.87	18.24
67	7,787.0	33.5	0.0000	0.548	0.001202	40.09	34.85	18.37
68	7,737.0	33.5	0.0000	0.553	0.001220	40.16	34.84	18.51

TABLE 3.12 EGLA Calculations for the Building in Example 3.2 (*Continued*)

Point	Distance from Shoreline, x_i (ft)	Transect Elevation, z_i (ft)	Topographic Slope, φ	Froude Number, F_n	Friction Slope, s_i	Energy Head, E_g (ft)	Inundation Depth, h_i (ft)	Flow Velocity, u_i (ft/s)
69	7,687.0	33.5	0.0000	0.557	0.001238	40.22	34.82	18.64
70	7,637.0	32.5	0.0200	0.561	0.001256	41.28	35.67	19.00
71	7,587.0	32.5	0.0000	0.565	0.001264	41.34	35.66	19.13
72	7,537.0	32.5	0.0000	0.569	0.001282	41.41	35.65	19.26
73	7,487.0	32.5	0.0000	0.572	0.001300	41.47	35.63	19.39
74	7,437.0	32.5	0.0000	0.576	0.001318	41.54	35.62	19.52
75	7,387.0	31.5	0.0200	0.580	0.001336	42.61	36.47	19.88
76	7,337.0	31.5	0.0000	0.584	0.001343	42.67	36.45	20.01
77	7,287.0	31.5	0.0000	0.588	0.001361	42.74	36.44	20.14
78	7,237.0	30.5	0.0200	0.592	0.001379	43.81	37.28	20.50
79	7,187.0	30.5	0.0000	0.596	0.001386	43.88	37.27	20.63
80	7,137.0	29.5	0.0200	0.599	0.001404	44.95	38.11	20.99
81	7,087.0	29.5	0.0000	0.603	0.001411	45.02	38.09	21.12
82	7,037.0	29.5	0.0000	0.607	0.001428	45.09	38.08	21.25
83	6,987.0	29.5	0.0000	0.610	0.001446	45.16	38.07	21.37
84	6,937.0	28.5	0.0200	0.614	0.001463	46.24	38.90	21.73
85	6,887.0	28.5	0.0000	0.618	0.001470	46.31	38.89	21.86
86	6,837.0	28.5	0.0000	0.621	0.001488	46.38	38.88	21.99
87	6,787.0	28.5	0.0000	0.625	0.001505	46.46	38.87	22.11
88	6,737.0	27.5	0.0200	0.629	0.001522	47.54	39.69	22.47
89	6,687.0	27.5	0.0000	0.632	0.001529	47.61	39.68	22.60
90	6,637.0	26.5	0.0200	0.636	0.001546	48.69	40.51	22.96
91	6,587.0	26.5	0.0000	0.639	0.001553	48.77	40.50	23.08
92	6,537.0	26.5	0.0000	0.643	0.001570	48.85	40.48	23.20
93	6,487.0	26.5	0.0000	0.646	0.001587	48.93	40.48	23.33
94	6,437.0	26.5	0.0000	0.650	0.001604	49.01	40.47	23.45
95	6,387.0	25.5	0.0200	0.653	0.001622	50.09	41.28	23.81
96	6,337.0	25.5	0.0000	0.657	0.001628	50.17	41.27	23.93
97	6,287.0	25.5	0.0000	0.660	0.001645	50.25	41.26	24.06
98	6,237.0	25.5	0.0000	0.663	0.001662	50.33	41.26	24.18
99	6,187.0	24.5	0.0200	0.667	0.001679	51.42	42.07	24.54
100	6,137.0	24.5	0.0000	0.670	0.001685	51.50	42.06	24.66
101	6,087.0	24.5	0.0000	0.673	0.001702	51.59	42.05	24.78
102	6,037.0	24.5	0.0000	0.677	0.001719	51.67	42.05	24.90
103	5,987.0	24.5	0.0000	0.680	0.001736	51.76	42.04	25.02

TABLE 3.12 EGLA Calculations for the Building in Example 3.2 (Continued)

Point	Distance from Shoreline, x_i (ft)	Transect Elevation, z_i (ft)	Topographic Slope, φ	Froude Number, F_n	Friction Slope, s_i	Energy Head, E_g (ft)	Inundation Depth, h_i (ft)	Flow Velocity, u_i (ft/s)
104	5,937.0	24.5	0.0000	0.683	0.001753	51.85	42.03	25.14
105	5,887.0	24.5	0.0000	0.687	0.001770	51.94	42.03	25.26
106	5,837.0	24.5	0.0000	0.690	0.001787	52.02	42.03	25.38
107	5,787.0	24.5	0.0000	0.693	0.001804	52.11	42.02	25.50
108	5,737.0	24.5	0.0000	0.696	0.001821	52.21	42.02	25.61
109	5,687.0	24.5	0.0000	0.700	0.001837	52.30	42.02	25.73
110	5,637.0	24.5	0.0000	0.703	0.001854	52.39	42.02	25.85
111	5,587.0	23.5	0.0200	0.706	0.001871	53.48	42.82	26.21
112	5,537.0	23.5	0.0000	0.709	0.001876	53.58	42.81	26.33
113	5,487.0	23.5	0.0000	0.712	0.001893	53.67	42.81	26.45
114	5,437.0	24.5	−0.0200	0.715	0.001910	52.77	42.02	26.31
115	5,387.0	24.5	0.0000	0.719	0.001939	52.86	42.02	26.43
116	5,337.0	24.5	0.0000	0.722	0.001955	52.96	42.02	26.55
117	5,287.0	24.5	0.0000	0.725	0.001972	53.06	42.02	26.66
118	5,237.0	24.5	0.0000	0.728	0.001989	53.16	42.03	26.78
119	5,187.0	24.5	0.0000	0.731	0.002006	53.26	42.03	26.89
120	5,137.0	24.5	0.0000	0.734	0.002023	53.36	42.04	27.00
121	5,087.0	24.5	0.0000	0.737	0.002039	53.46	42.04	27.12
122	5,037.0	23.5	0.0200	0.740	0.002056	54.57	42.84	27.49
123	4,987.0	23.5	0.0000	0.743	0.002060	54.67	42.84	27.60
124	4,937.0	23.5	0.0000	0.746	0.002077	54.77	42.85	27.71
125	4,887.0	24.5	−0.0200	0.749	0.002093	53.88	42.07	27.57
126	4,837.0	24.5	0.0000	0.752	0.002123	53.98	42.08	27.69
127	4,787.0	24.5	0.0000	0.755	0.002140	54.09	42.09	27.80
128	4,737.0	23.5	0.0200	0.758	0.002156	55.20	42.88	28.17
129	4,687.0	23.5	0.0000	0.761	0.002160	55.31	42.89	28.28
130	4,637.0	23.5	0.0000	0.764	0.002177	55.42	42.90	28.39
131	4,587.0	23.5	0.0000	0.767	0.002193	55.53	42.91	28.51
132	4,537.0	23.5	0.0000	0.770	0.002210	55.64	42.92	28.62
133	4,487.0	22.5	0.0200	0.773	0.002226	56.75	43.70	28.99
134	4,437.0	22.5	0.0000	0.776	0.002229	56.86	43.71	29.10
135	4,387.0	22.5	0.0000	0.779	0.002246	56.97	43.72	29.21
136	4,337.0	21.5	0.0200	0.781	0.002262	58.08	44.50	29.58
137	4,287.0	21.5	0.0000	0.784	0.002266	58.20	44.51	29.69
138	4,237.0	20.5	0.0200	0.787	0.002282	59.31	45.28	30.06

TABLE 3.12   EGLA Calculations for the Building in Example 3.2 (Continued)

Point	Distance from Shoreline, x_i (ft)	Transect Elevation, z_i (ft)	Topographic Slope, φ	Froude Number, F_n	Friction Slope, s_i	Energy Head, E_g (ft)	Inundation Depth, h_i (ft)	Flow Velocity, u_i (ft/s)
139	4,187.0	20.5	0.0000	0.790	0.002285	59.43	45.29	30.17
140	4,137.0	20.5	0.0000	0.793	0.002302	59.54	45.30	30.28
141	4,087.0	20.5	0.0000	0.796	0.002318	59.66	45.31	30.39
142	4,037.0	19.5	0.0200	0.798	0.002334	60.77	46.08	30.76
143	3,987.0	19.5	0.0000	0.801	0.002337	60.89	46.09	30.87
144	3,937.0	19.5	0.0000	0.804	0.002354	61.01	46.11	30.98
145	3,887.0	18.5	0.0200	0.807	0.002370	62.13	46.87	31.34
146	3,837.0	18.5	0.0000	0.810	0.002373	62.25	46.88	31.46
147	3,787.0	17.5	0.0200	0.812	0.002389	63.37	47.64	31.82
148	3,737.0	17.5	0.0000	0.815	0.002393	63.48	47.65	31.93
149	3,687.0	17.5	0.0000	0.818	0.002409	63.61	47.66	32.04
150	3,637.0	17.5	0.0000	0.821	0.002425	63.73	47.67	32.15
151	3,587.0	17.5	0.0000	0.823	0.002440	63.85	47.69	32.26
152	3,537.0	17.5	0.0000	0.826	0.002456	63.97	47.70	32.37
153	3,487.0	16.5	0.0200	0.829	0.002472	65.09	48.45	32.74
154	3,437.0	16.5	0.0000	0.831	0.002476	65.22	48.46	32.85
155	3,387.0	16.5	0.0000	0.834	0.002491	65.34	48.48	32.96
156	3,337.0	16.5	0.0000	0.837	0.002507	65.47	48.49	33.07
157	3,287.0	16.5	0.0000	0.840	0.002523	65.59	48.50	33.18
158	3,237.0	15.5	0.0200	0.842	0.002539	66.72	49.25	33.54
159	3,187.0	15.5	0.0000	0.845	0.002542	66.85	49.27	33.65
160	3,137.0	15.5	0.0000	0.848	0.002558	66.98	49.28	33.76
161	3,087.0	15.5	0.0000	0.850	0.002574	67.11	49.29	33.87
162	3,037.0	15.5	0.0000	0.853	0.002589	67.23	49.30	33.98
163	2,987.0	15.5	0.0000	0.855	0.002605	67.37	49.32	34.09
164	2,937.0	15.5	0.0000	0.858	0.002621	67.50	49.33	34.20
165	2,887.0	15.5	0.0000	0.861	0.002637	67.63	49.35	34.31
166	2,837.0	14.5	0.0200	0.863	0.002652	68.76	50.09	34.67
167	2,787.0	14.5	0.0000	0.866	0.002655	68.89	50.11	34.78
168	2,737.0	14.5	0.0000	0.868	0.002671	69.03	50.12	34.89
169	2,687.0	14.5	0.0000	0.871	0.002686	69.16	50.14	35.00
170	2,637.0	13.5	0.0200	0.874	0.002702	70.30	50.88	35.36
171	2,587.0	13.5	0.0000	0.876	0.002704	70.43	50.90	35.47
172	2,537.0	13.5	0.0000	0.879	0.002720	70.57	50.91	35.58
173	2,487.0	13.5	0.0000	0.881	0.002736	70.70	50.93	35.69

TABLE 3.12 EGLA Calculations for the Building in Example 3.2 (*Continued*)

Point	Distance from Shoreline, x_i (ft)	Transect Elevation, z_i (ft)	Topographic Slope, φ	Froude Number, F_n	Friction Slope, s_i	Energy Head, E_g (ft)	Inundation Depth, h_i (ft)	Flow Velocity, u_i (ft/s)
174	2,437.0	13.5	0.0000	0.884	0.002751	70.84	50.94	35.80
175	2,387.0	14.5	−0.0200	0.886	0.002767	69.98	50.24	35.65
176	2,337.0	14.5	0.0000	0.889	0.002795	70.12	50.26	35.76
177	2,287.0	13.5	0.0200	0.891	0.002811	71.26	51.00	36.12
178	2,237.0	13.5	0.0000	0.894	0.002813	71.40	51.02	36.23
179	2,187.0	13.5	0.0000	0.896	0.002828	71.54	51.04	36.34
180	2,137.0	13.5	0.0000	0.899	0.002844	71.68	51.06	36.45
181	2,087.0	13.5	0.0000	0.901	0.002859	71.83	51.08	36.56
182	2,037.0	13.5	0.0000	0.904	0.002875	71.97	51.10	36.67
183	1,987.0	13.5	0.0000	0.906	0.002890	72.12	51.12	36.77
184	1,937.0	13.5	0.0000	0.909	0.002906	72.26	51.14	36.88
185	1,887.0	13.5	0.0000	0.911	0.002921	72.41	51.16	36.99
186	1,837.0	14.5	−0.0200	0.914	0.002936	71.55	50.48	36.84
187	1,787.0	14.5	0.0000	0.916	0.002965	71.70	50.50	36.95
188	1,737.0	13.5	0.0200	0.919	0.002981	72.85	51.23	37.31
189	1,687.0	13.5	0.0000	0.921	0.002982	73.00	51.25	37.42
190	1,637.0	13.5	0.0000	0.924	0.002998	73.15	51.28	37.53
191	1,587.0	13.5	0.0000	0.926	0.003013	73.30	51.30	37.64
192	1,537.0	13.5	0.0000	0.928	0.003028	73.45	51.33	37.75
193	1,487.0	14.5	−0.0200	0.931	0.003043	72.60	50.66	37.59
194	1,437.0	14.5	0.0000	0.933	0.003073	72.76	50.69	37.70
195	1,387.0	14.5	0.0000	0.936	0.003088	72.91	50.71	37.81
196	1,337.0	14.5	0.0000	0.938	0.003103	73.07	50.74	37.92
197	1,287.0	14.5	0.0000	0.940	0.003118	73.22	50.77	38.03
198	1,237.0	14.5	0.0000	0.943	0.003134	73.38	50.80	38.13
199	1,187.0	14.5	0.0000	0.945	0.003149	73.54	50.83	38.24
200	1,137.0	13.5	0.0200	0.948	0.003164	74.70	51.55	38.61
201	1,087.0	13.5	0.0000	0.950	0.003165	74.85	51.58	38.71
202	1,037.0	13.5	0.0000	0.952	0.003180	75.01	51.61	38.82
203	987.0	13.5	0.0000	0.955	0.003195	75.17	51.64	38.93
204	937.0	13.5	0.0000	0.957	0.003210	75.33	51.67	39.04
205	887.0	14.5	−0.0200	0.959	0.003226	74.49	51.02	38.88
206	837.0	14.5	0.0000	0.962	0.003255	74.66	51.05	38.99
207	787.0	14.5	0.0000	0.964	0.003270	74.82	51.08	39.10
208	737.0	15.5	−0.0200	0.966	0.003285	73.99	50.44	38.94
209	687.0	15.5	0.0000	0.969	0.003315	74.15	50.47	39.05

Tᴀʙʟᴇ **3.12** EGLA Calculations for the Building in Example 3.2 (*Continued*)

Point	Distance from Shoreline, x_i (ft)	Transect Elevation, z_i (ft)	Topographic Slope, φ	Froude Number, F_n	Friction Slope, s_i	Energy Head, E_g (ft)	Inundation Depth, h_i (ft)	Flow Velocity, u_i (ft/s)
210	637.0	15.5	0.0000	0.971	0.003330	74.32	50.51	39.16
211	587.0	16.5	−0.0200	0.973	0.003345	73.48	49.87	39.00
212	537.0	15.5	0.0200	0.976	0.003375	74.65	50.58	39.37
213	487.0	15.5	0.0000	0.978	0.003375	74.82	50.62	39.48
214	437.0	15.5	0.0000	0.980	0.003390	74.99	50.66	39.59
215	387.0	15.5	0.0000	0.982	0.003405	75.16	50.70	39.69
216	337.0	15.5	0.0000	0.985	0.003420	75.33	50.73	39.80
217	287.0	14.5	0.0200	0.987	0.003435	76.50	51.45	40.17
218	237.0	13.5	0.0200	0.989	0.003435	77.68	52.15	40.54
219	187.0	12.5	0.0200	0.992	0.003435	78.85	52.86	40.91
220	137.0	10.5	0.0400	0.994	0.003435	81.02	54.24	41.53
221	87.0	9.5	0.0200	0.996	0.003422	82.19	54.94	41.89
222	37.0	8.5	0.0200	0.998	0.003422	83.36	55.64	42.26
223	0.0	7.5	0.0270	1.000	0.003419	84.49	56.33	42.59

1 ft = 0.3048 m; 1 ft/s = 0.31 m/s

TABLE 3.12 EGLA Calculations for the Building in Example 3.2 (*Continued*)

Step 4a—Acquire x_R and R

$R = 70.0$ ft (21.3 m) is obtained from an R/H_T analysis (see Step 3).
R occurs at $x_R = 11,137$ ft (3,395 m) from the shoreline.

Step 4b—Approximate the transect by a series of segmented slopes with transect points spaced not more than 100 ft (30.5 m).

The ground transect at this location is given in Fig. 3.33 with a maximum transect point spacing of 50 ft (15.2 m).

Step 4c—Calculate the slope within the segment

$$\varphi_{137} = \frac{z_{136} - z_{137}}{x_{136} - x_{137}} = \frac{21.5 - 21.5}{4,337.0 - 4,287.0} = 0.0000$$

Step 4d—Obtain Manning's coefficient, n, from ASCE/SEI Table 6.6-1 for the segment

Assume the frictional surface for this segment (as well as all the other segments in the transect) does not fall under any specific description in ASCE/SEI Table 6.6-1. Therefore, $n = 0.03$.

Step 4e—Calculate the Froude number, F_{r137}

$$F_{r137} = \alpha\left(1 - \frac{x_{137}}{x_R}\right)^{0.5} = 1.0 \times \left(1 - \frac{4,287.0}{11,137.0}\right)^{0.5} = 0.7843$$

Step 4f—Calculate the hydraulic friction slope, s_{137}

$$s_{137} = \frac{gF_{r137}^2}{\left(\dfrac{1.49}{n}\right)^2 h_{136}^{1/3}} = \frac{32.2 \times 0.7843^2}{\left(\dfrac{1.49}{0.03}\right)^2 \times (44.50)^{1/3}} = 0.002266$$

In S.I.:

$$s_{137} = \frac{gF_{r137}^2}{\left(\dfrac{1.00}{n}\right)^2 h_{136}^{1/3}} = \frac{9.81 \times 0.7843^2}{\left(\dfrac{1.00}{0.03}\right)^2 \times (13.56)^{1/3}} = 0.002277$$

Step 4g—Calculate the hydraulic head, $E_{g,137}$

$$E_{g,137} = E_{g,136} + (\varphi_{137} + s_{137})(x_{136} - x_{137})$$

$$= 58.08 + [(0.0000 + 0.002266) \times 50] = 58.20 \text{ ft}$$

In S.I.:

$$E_{g,1137} = 17.70 + [(0.0000 + 0.002277) \times 15.2] = 17.74 \text{ m}$$

Step 4h—Calculate the inundation depth, h_{137}

$$h_{137} = \frac{E_{g,137}}{1 + 0.5F_{r137}^2} = \frac{58.20}{1 + (0.5 \times 0.7843^2)} = 44.51 \text{ ft}$$

In S.I.:

$$h_{137} = \frac{E_{g,137}}{1 + 0.5F_{r137}^2} = \frac{17.74}{1 + (0.5 \times 0.7843^2)} = 13.57 \text{ m}$$

Step 4i—Calculate the flow velocity, u_{137}

$$u_{137} = F_{r,137}(gh_{137})^{0.5} = 0.7843 \times (32.2 \times 44.51)^{0.5} = 29.69 \text{ ft/s}$$

$$> 10.00 \text{ ft/s}$$

$$< \text{lesser of} \begin{cases} 1.5(gh_{137})^{0.5} = 56.79 \text{ ft/s} \\ 50.00 \text{ ft/s} \end{cases}$$

In S.I.:

$$u_{137} = F_{r,137}(gh_{137})^{0.5} = 0.7843 \times (9.81 \times 13.57)^{0.5} = 9.05 \text{ m/s}$$

$$> 3.05 \text{ ft/s}$$

$$< \text{lesser of} \begin{cases} 1.5(gh_{137})^{0.5} = 17.31 \text{ m/s} \\ 15.24 \text{ m/s} \end{cases}$$

Plots of elevation, inundation depth, and flow velocity along the transect are given in Fig. 3.34.

The effects of sea level change in accordance with ASCE/SEI 6.5.3 are not considered in this example but should be included wherever appropriate.

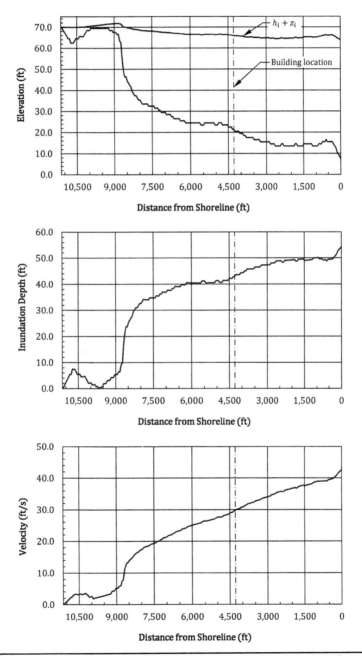

FIGURE 3.34   Elevation, inundation depth, and flow velocity along the transect in Example 3.2.

Additionally, it is assumed the flow velocity amplification requirements in ASCE/SEI 6.8.5 are not applicable.

Similar analyses must be performed for transects located 22.5 degrees on either side of the transect in this example (see ASCE/SEI 6.8.6.1 and Fig. 3.7).

Step 5—Determine the hydrostatic loads   Sec. 3.9

Step 5a—Buoyancy loads

Because the slab-on-grade is isolated from structure, buoyancy loads on the overall building need not be considered.

Step 5b—Unbalanced lateral hydrostatic loads

This load is applicable to the inundated exterior walls on column lines A and F because they are solid, are greater than 30 ft (9.1 m) in length, and are not adjacent to tsunami breakaway walls. This load occurs during the load case 1 and load case 2 inflow cases.

Load case 1:

$$h_{sx} = \text{lesser of} \begin{cases} h_{max} = 44.5 \ (13.0 \ \text{m}) \\ \text{Story height} = 11.25 \ \text{ft} \ (3.4 \ \text{m}) \\ \text{Height to top of first story window} = 11.25 - 1.0 = 10.3 \ \text{ft} \ (3.1 \ \text{m}) \end{cases}$$

$$F_h = \frac{1}{2}\gamma_s b h_{sx}^2 = \frac{1}{2} \times (1.1 \times 64.0) \times 32.0 \times 10.3^2/1,000 = 120 \ \text{kips}$$

In S.I.: $F_h = \frac{1}{2} \times (1.1 \times 10.05) \times 9.8 \times 3.1^2 = 521 \ \text{kN}$

Load case 2:

$$h_{max} = \frac{2}{3} \times 44.5 = 29.7 \ \text{ft} \ (9.0 \ \text{m})$$

$$F_h = \frac{1}{2} \times (1.1 \times 64.0) \times 32.0 \times 29.7^2/1,000 = 994 \ \text{kips}$$

In S.I.: $F_h = \frac{1}{2} \times (1.1 \times 10.05) \times 9.8 \times 9.0^2 = 4,388 \ \text{kN}$

The unbalanced hydrostatic load, F_h, is illustrated in Fig. 3.35 for load case 2.

Step 5c—Residual water surcharge load on floors and walls

Because $h_{max} = 44.5$ ft (13.0 m), the bottom three floors are inundated. The segments of the perimeter upturned beams above the floor slab retain water on these floors, so the residual water surcharge load on the floor slabs, p_r, is equal to the following (see Fig. 3.10 and Detail A-A in Fig. 3.32):

$$p_r = \gamma_s h_r = (1.1 \times 64.0) \times (18.0/12) = 106 \ \text{lb/ft}^2$$

In S.I.:

$$p_r = (1.1 \times 10.05) \times 0.46 = 5.1 \ \text{kN/m}^2$$

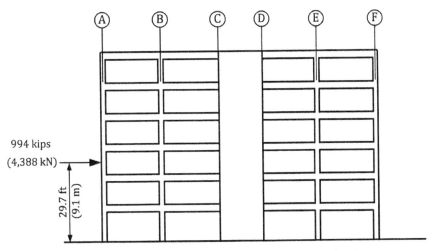

994 kips
(4,388 kN)

29.7 ft
(9.1 m)

This force acts on the wall on column line A (or F)

FIGURE 3.35 Unbalanced lateral hydrostatic force on the exterior walls in Example 3.2.

The 18-in. (457-mm) wide by 18-in. (457-mm) deep segment of the upturned beam is subjected to a triangular lateral load with a maximum pressure of 106 lb/ft² (5.1 kN/m²) at the base of the segment (see Fig. 3.36).

Step 5d—Hydrostatic surcharge pressure on foundation

As noted in the design data, the building is supported by piles, so this pressure is not applicable in this case.

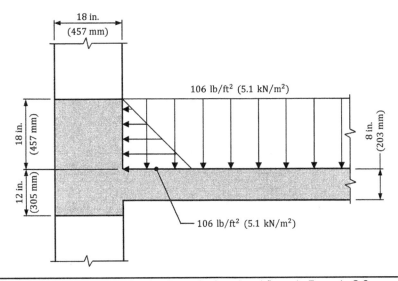

18 in.
(457 mm)

106 lb/ft² (5.1 kN/m²)

18 in.
(457 mm)

12 in.
(305 mm)

8 in.
(203 mm)

106 lb/ft² (5.1 kN/m²)

FIGURE 3.36 Residual water surcharge loads on the inundated floors in Example 3.2.

Step 6—Determine the hydrodynamic loads   Sec. 3.10

Step 6a—Option 1: Simplified uniform lateral static pressure

$$p_{uw} = 1.25 I_{tsu} \gamma_s h_{max} = 1.25 \times 1.0 \times (1.1 \times 64.0) \times 44.5 = 3,916 \text{ lb/ft}^2 \qquad (3.8)$$

This pressure is applied over the width of the building and over a height of $1.3 h_{max} = 57.9$ ft

In S.I.:

$$p_{uw} = 1.25 I_{tsu} \gamma_s h_{max} = 1.25 \times 1.0 \times (1.1 \times 10.05) \times 13.6 = 187.9 \text{ kN/m}^2$$

This pressure is applied over the width of the building and over a height of $1.3 h_{max} = 17.7$ m.

$$\text{Total force} = 3,916 \times 183.5 \times 57.9 / 1,000 = 41,606 \text{ kips}$$

In S.I.:

$$\text{Total force} = 187.9 \times 55.9 \times 17.7 = 185,914 \text{ kN}$$

The simplified uniform lateral static pressure is depicted in Fig. 3.37.

Step 6b—Option 2: Detailed hydrodynamic lateral forces   Table 3.5

• Overall drag force on building

The flowchart in Fig. 3.13 is used to determine F_{dx} for load cases 1, 2, and 3.

$$\rho_s = 2.2 \text{ slugs/ft}^3 \ (1,128 \text{ kg/m}^3)$$

For Tsunami Category II buildings, $I_{tsu} = 1.0$ from ASCE/SEI Table 6.8-1. Values of C_d for load cases 1, 2, and 3 are given in Table 3.13. Sample calculations for load cases 1 and 2 are as follows:

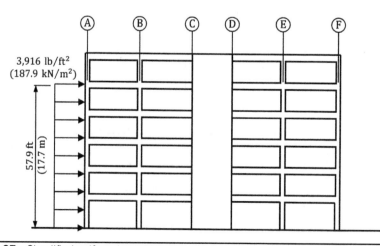

Figure 3.37   Simplified uniform lateral static hydrodynamic pressure in Example 3.2.

Load Case	h_{sx}, ft (m)	B/h_{sx}	C_d
1	10.3 (3.1)	17.8	1.32
2	29.7 (9.1)	6.2	1.25
3	44.5 (13.6)	4.1	1.25

TABLE 3.13 Values of C_d for Load Cases 1, 2, and 3 in Example 3.2

For load case 1:

$$h_{sx} = \text{lesser of} \begin{cases} h_{max} = 44.5 \ (13.0 \text{ m}) \\ \text{Story height} = 11.25 \text{ ft } (3.4 \text{ m}) \\ \text{Height to top of first story window} = 11.25 - 1.0 = 10.3 \text{ ft } (3.1 \text{ m}) \end{cases}$$

$$B/h_{sx} = 183.5/10.3 = 17.8$$

From ASCE/SEI Table 6.10-1, $C_d = 1.32$ by linear interpolation
For load case 2:

$$h_{sx} = 2h_{max}/3 = 29.7 \text{ ft } (9.1 \text{ m})$$

$$B/h_{sx} = 183.5/29.7 = 6.2$$

From ASCE/SEI Table 6.10-1, $C_d = 1.25$

Values of C_{cx} for load cases 1, 2, and 3 at a typical floor are given in Table 3.14. Sample calculations for load cases 1 and 2 are as follows:
For load case 1:

$C_{cx} = 1.0$ because it is assumed the exterior walls have not failed

For load case 2:

$$h_{sx} = 2h_{max}/3 = 29.7 \text{ ft } (9.1 \text{ m})$$

Edge columns: $A_{col} = (18.0/12) \times 29.7 = 44.6 \text{ ft}^2 \ (4.1 \text{ m}^2)$
Interior columns: $A_{col} = (20.0/12) \times 29.7 = 49.5 \text{ ft}^2 \ (4.6 \text{ m}^2)$
Walls around floor openings: $A_{wall} = 10.0 \times 29.7 = 297.0 \text{ ft}^2 \ (27.6 \text{ m}^2)$

Load Case	h_{sx}, ft (m)	C_{cx}
1	10.3 (3.1)	1.0
2	29.7 (9.1)	1.0
3	44.5 (13.6)	1.0

TABLE 3.14 Values of C_{cx} for Load Cases 1, 2, and 3 in Example 3.2

Walls on column lines A and F: $A_{wall} = 32.0 \times 29.7 = 950.4$ ft^2 (88.3 m^2)

$$A_{beam} = (30.0/12) \times (183.5 - 32.0) = 378.8 \text{ ft}^2 \ (35.2 \text{ m}^2)$$

$$C_{cx} = \frac{\sum(A_{col} + A_{wall}) + 1.5A_{beam}}{Bh_{sx}}$$

$$= \frac{[(16 \times 44.6) + (24 \times 49.5)] + [(2 \times 297.0) + (2 \times 950.4)] + (2 \times 1.5 \times 378.8)}{183.5 \times 29.7}$$

$$= 1.02 > 1.0, \text{ use } 1.0$$

In S.I.:

$$C_{cx} = \frac{\sum(A_{col} + A_{wall}) + 1.5A_{beam}}{Bh_{sx}}$$

$$= \frac{[(16 \times 4.1) + (24 \times 4.6)] + [(2 \times 27.6) + (2 \times 88.3)] + (2 \times 1.5 \times 35.2)}{55.9 \times 9.1}$$

$$= 1.01 > 1.0, \text{ use } 1.0$$

For load case 1, determine the flow velocity from ASCE/SEI Figure 6.8-1. With $h/h_{max} = 10.3/44.5 = 0.23$, $u/u_{max} \cong 0.7$. Therefore, use $u = 0.7 \times 29.7 = 20.8$ ft/s (6.3 m/s).

The overall drag force, F_{dx}, for load case 1 is as follows:

$$F_{dx} = \frac{1}{2}\rho_s I_{tsu} C_d C_{cx} B(hu^2)$$

$$= \frac{1}{2} \times 2.2 \times 1.0 \times 1.32 \times 1.0 \times 183.5 \times (10.3 \times 20.8^2)/1,000 = 1,187 \text{ kips}$$

In S.I.:

$$F_{dx} = \frac{1}{2} \times (1,128/1,000) \times 1.0 \times 1.32 \times 1.0 \times 55.9 \times (3.1 \times 6.3^2) = 5,120 \text{ kN}$$

The hydrodynamic loads in load case 1 are shown in Fig. 3.38.

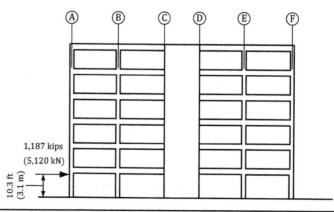

FIGURE 3.38 Load case 1 for the building in Example 3.2.

For load case 2:

$$F_{dx} = \frac{1}{2}\rho_s I_{tsu} C_d C_{cx} B(hu^2)$$

$$= \frac{1}{2} \times 2.2 \times 1.0 \times 1.25 \times 1.0 \times 183.5 \times (29.7 \times 29.7^2)/1,000 = 6,610 \text{ kips}$$

In S.I.:

$$F_{dx} = \frac{1}{2} \times (1,128/1,000) \times 1.0 \times 1.25 \times 1.0 \times 55.9 \times (9.1 \times 9.1^2) = 29,698 \text{ kN}$$

This force is applied to the two inundated floor levels based on the tributary height, as shown in Fig. 3.39.

For load case 3, $u = u_{max}/3 = 29.7/3 = 9.9$ ft/s (3.0 m/s) < 10.0 ft/s (3.1 m/s) and $h = h_{max} = 44.5$ ft (13.6 m). Therefore, with $u = 10.0$ ft/s (3.1 m/s), $C_d = 1.25$, and $C_{cx} = 1.0$, $F_{dx} = 1,123$ kips (5,151 kN). This force is applied to the three inundated floor levels based on the tributary height, as shown in Fig. 3.40.

- Drag force on components

The flowchart in Fig. 3.15 is used to determine F_d for the interior and exterior components.

$$\rho_s = 2.2 \text{ slugs/ft}^3 \ (1,128 \text{ kg/m}^3)$$

For Tsunami Category II buildings, $I_{tsu} = 1.0$ from ASCE/SEI Table 6.8-1.

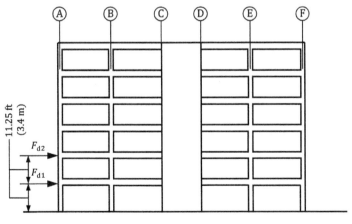

$$F_{d1} = \frac{11.25 + \frac{11.25}{2}}{29.7} \times 6,610 = 3,756 \text{ kips (16,874 kN)}$$

$$F_{d2} = \frac{\frac{11.25}{2} + (29.7 - 22.5)}{29.7} \times 6,610 = 2,854 \text{ kips (12,824 kN)}$$

FIGURE 3.39 Load case 2 for the building in Example 3.2.

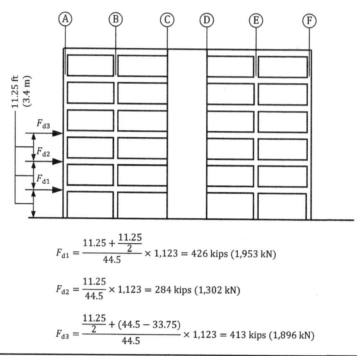

$$F_{d1} = \frac{11.25 + \frac{11.25}{2}}{44.5} \times 1{,}123 = 426 \text{ kips (1,953 kN)}$$

$$F_{d2} = \frac{11.25}{44.5} \times 1{,}123 = 284 \text{ kips (1,302 kN)}$$

$$F_{d3} = \frac{\frac{11.25}{2} + (44.5 - 33.75)}{44.5} \times 1{,}123 = 413 \text{ kips (1,896 kN)}$$

FIGURE 3.40   Load case 3 for the building in Example 3.2.

Interior columns:

For square columns, $C_d = 2.0$ from ASCE/SEI Table 6.10-2.

$$b = 20.0/12 = 1.7 \text{ ft (0.51 m)}$$

Load case 2 governs where inundated height $h_e = 11.25 - 1.0 = 10.25$ ft (3.1 m) and $u = u_{max} = 29.7$ ft/s (9.1 m/s):

$$F_d = \frac{1}{2}\rho_s I_{tsu} C_d b(h_e u^2) \tag{3.10}$$

$$= \frac{1}{2} \times 2.2 \times 1.0 \times 2.0 \times 1.7 \times (10.25 \times 29.7^2)/1{,}000 = 34 \text{ kips}$$

This load is applied as an equivalent uniformly distributed lateral load of $34/10.25 = 3.3$ kips/ft over the height of the column.

In S.I:

$$F_d = \frac{1}{2} \times (1{,}128/1{,}000) \times 1.0 \times 2.0 \times 0.51 \times (3.1 \times 9.1^2) = 148 \text{ kN}$$

This load is applied as an equivalent uniformly distributed lateral load of $148/3.1 = 47.7$ kN/m over the height of the column.

Exterior columns:

$$C_d = 2.0 \qquad\qquad \text{ASCE/SEI 6.10.2.2}$$

Tributary width of an exterior column = 25.0 ft (7.6 m)

$$C_{cx} = 0.7$$
$$b = C_{cx} \times \text{Tributary width} = 0.7 \times 25.0 = 17.5 \text{ ft (5.3 m)}$$

Load case 2 governs where inundated height $h_e = 10.25$ ft (3.1 m) and $u = u_{max} = 29.7$ ft/s (9.1 m/s):

$$F_d = \frac{1}{2}\rho_s I_{tsu} C_d b (h_e u^2) \qquad\qquad (3.10)$$

$$= \frac{1}{2} \times 2.2 \times 1.0 \times 2.0 \times 17.5 \times (10.25 \times 29.7^2)/1,000 = 348 \text{ kips}$$

This load is applied as an equivalent uniformly distributed lateral load of $348/10.25 = 34.0$ kips/ft over the height of the column.

In S.I:

$$F_d = \frac{1}{2} \times (1,128/1,000) \times 1.0 \times 2.0 \times 5.3 \times (3.1 \times 9.1^2) = 1,535 \text{ kN}$$

This load is applied as an equivalent uniformly distributed lateral load of $1,535/3.1 = 495.2$ kN/m over the height of the column.

Interior walls:

For a wall normal to the flow, $C_d = 2.0$ from ASCE/SEI Table 6.10-2.

$$b = 10.0 \text{ ft (3.1 m)}$$

Load case 2 governs where inundated height $h_e = 11.25$ ft (3.4 m) and $u = u_{max} = 29.7$ ft/s (9.1m/s):

$$F_d = \frac{1}{2}\rho_s I_{tsu} C_d b (h_e u^2) \qquad\qquad (3.10)$$

$$= \frac{1}{2} \times 2.2 \times 1.0 \times 2.0 \times 10.0 \times (11.25 \times 29.7^2)/1,000 = 218 \text{ kips}$$

This load is applied as an equivalent uniformly distributed lateral load of $218/11.25 = 19.4$ kips/ft over the height of the wall.

In S.I:

$$F_d = \frac{1}{2} \times (1,128/1,000) \times 1.0 \times 2.0 \times 3.1 \times (3.4 \times 9.1^2) = 985 \text{ kN}$$

This load is applied as an equivalent uniformly distributed lateral load of $985/3.4 = 289.7$ kN/m over the height of the wall.

Exterior walls:

For a wall normal to the flow, $C_d = 2.0$ from ASCE/SEI Table 6.10-2.

$$b = 32.0 \text{ ft (9.8 m)}$$

Load case 2 governs where inundated height $h_e = 11.25$ ft (3.4 m) and $u = u_{max} = 29.7$ ft/s (9.1 m/s):

$$F_d = \frac{1}{2}\rho_s I_{tsu} C_d b(h_e u^2) \tag{3.10}$$

$$= \frac{1}{2}\times 2.2\times 1.0\times 2.0\times 32.0\times(11.25\times 29.7^2)/1,000 = 699 \text{ kips}$$

This load is applied as an equivalent uniformly distributed lateral load of $699/11.25 = 62.1$ kips/ft over the height of the wall.

In S.I:

$$F_d = \frac{1}{2}\times(1,128/1,000)\times 1.0\times 2.0\times 9.8\times(3.4\times 9.1^2) = 3,112 \text{ kN}$$

This load is applied as an equivalent uniformly distributed lateral load of $3,112/3.4 = 915.3$ kN/m over the height of the wall.

• Tsunami loads on vertical structural components

The exterior walls are subjected to tsunami bore hydrodynamic loads.

An EGLA was performed using Froude number coefficient $\alpha = 1.3$ in accordance with ASCE/SEI 6.6.2, which resulted in the following at the site:

$$F_n = 1.02$$

$$h_{max} = 42.6 \text{ ft (13.0 m)}$$

$$u_{max} = 37.8 \text{ ft/s (11.5 m/s)}$$

For a wall normal to the flow, $C_d = 2.0$ from ASCE/SEI Table 6.10-2.

$$b = 32.0 \text{ ft (9.8 m)}$$

Load case 2 governs where inundated height $h_e = 32.0/3 = 10.7$ ft (3.3 m) in accordance with ASCE/SEI 6.10.2.3 and $u = u_{max} = 37.8$ ft/s (11.5 m/s):

$$F_w = \frac{3}{4}\rho_s I_{tsu} C_d b(h_e u^2)_{bore} \qquad \text{ASCE/SEI Equation (6.10-5}b\text{)}$$

$$= \frac{3}{4}\times 2.2\times 1.0\times 2.0\times 32.0\times(10.7\times 37.8^2)/1,000 = 1,615 \text{ kips}$$

In S.I:

$$F_w = \frac{3}{4}\times(1,128/1,000)\times 1.0\times 2.0\times 9.8\times(3.3\times 11.5^2) = 7,237 \text{ kN}$$

This load is applied to the wall at points critical for flexure and shear.

- Hydrodynamic loads on perforated walls

  This load is not applicable in this example because there are no perforated walls.

- Walls angled to the flow

  This load is not applicable in this example because there are no walls angled to the flow.

- Hydrodynamic pressure associated with slabs

  The loads associated with (1) spaces in buildings subjected to flow stagnation pressurization, (2) hydrodynamic surge uplift at horizontal slabs, and (3) tsunami pore flow entrapped in structural wall-slab recess are not applicable in this example.

Step 7—Determine the debris impact loads   Sec. 3.11

Because the inundation depth exceeds 3 ft (0.91 m) at the site, the exterior columns and walls below the flow depth are subjected to debris impact loads in accordance with ASCE/SEI 6.11.

Step 7a—Option 1: Alternative simplified debris impact static load

$$F_i = 330C_oI_{tsu} = 330 \times 0.65 \times 1.0 = 215 \text{ kips} \tag{3.15a}$$

In S.I.:

$$F_i = 1,470 \ C_oI_{tsu} = 1,470 \times 0.65 \times 1.0 = 956 \text{ kN} \tag{3.15b}$$

Because the site is not located in an impact zone for shipping containers, it is permitted to reduce the calculated value of the impact force by 50 percent:

$$F_i = 0.5 \times 215 = 108 \text{ kips (478 kN)}$$

The impact force is applied to exterior columns and walls at points along the height of the component critical for flexure and shear. Illustrated in Fig. 3.41 is the impact force applied at $10.25/2 = 5.13$ ft (1.6 m) from the base of an exterior column, which is the critical section for flexure.

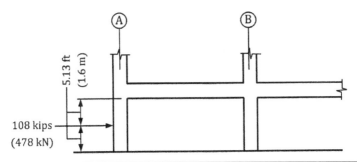

Figure 3.41 Alternative simplified debris impact static load on an exterior column in the building in Example 3.2.

Step 7b—Option 2: Detailed debris impact loads

- Wood logs and poles

The flowchart in Fig. 3.17 is used to determine the impact force due to wood logs and poles.

$$u_{max} = 29.7 \text{ ft/s (9.1 m/s)}$$

Minimum log stiffness = 350 kip/in. (61,300 kN/m)

Exterior column:

Assuming the column has simple connections at both ends, the stiffness of the column is determined as follows:

$$k = \frac{48EI}{L^3} = \frac{48 \times (3,605,000/1,000) \times \left(\frac{1}{12} \times 18^4\right)}{(10.25 \times 12)^3} = 814 \text{ kip/in.} > 350 \text{ kip/in.}$$

In S.I.:

$$k = \frac{48 \times (24,900 \times 10^3) \times \left[\frac{1}{12} \times \left(\frac{457}{1,000}\right)^4\right]}{(3.1)^3} = 145,827 \text{ kN/m} > 61,300 \text{ kN/m}$$

Therefore, use $k = 350$ kip/in. (61,300 kN/m).
Assume minimum debris weight $W_d = 1,000$ lb (4.5 kN).

Debris mass, m_d:

$$m_d = W_d/g = 1,000/32.2 = 31.1 \text{ slugs}$$

In S.I.:

$$m_d = 4.5 \times 10^3/9.81 = 459 \text{ kg}$$

Nominal debris impact force, F_{ni}:

$$F_{ni} = u_{max}\sqrt{km_d} = 29.7 \times \sqrt{(350 \times 12) \times (31.1/1,000)} = 339 \text{ kips}$$

In S.I.:

$$F_{ni} = 9.1 \times \sqrt{(61,300/1,000) \times 459} = 1,526 \text{ kN}$$

For Tsunami Category II buildings, $I_{tsu} = 1.0$ from ASCE/SEI Table 6.8-1.t

Orientation coefficient $C_o = 0.65$.

Design debris impact force $F_i = I_{tsu}C_oF_{ni} = 1.0 \times 0.65 \times 339 = 220 \text{ kips}$

In S.I.:

Design debris impact force $F_i = 1.0 \times 0.65 \times 1,526 = 992 \text{ kN}$

Impulse duration for elastic impact, t_d:

$$t_d = \frac{2m_d u_{max}}{F_{ni}} = \frac{2 \times 31.1 \times 29.7}{339 \times 1,000} = 0.0054 \text{ s}$$

In S.I.:

$$t_d = \frac{2m_d u_{max}}{F_{ni}} = \frac{2 \times 459 \times 9.1}{1,526 \times 1,000} = 0.0055 \text{ s}$$

Assuming the column has simple connections at both ends, the natural period of the column is determined as follows:

$$\text{Mass of column per unit length } m = \frac{18 \times 18}{144} \times \frac{150}{32.2} = 10.5 \text{ slugs/ft}$$

$$T_n = \frac{2L^2}{\pi\sqrt{\dfrac{EI}{m}}} = \frac{2 \times 10.25^2}{\pi\sqrt{\dfrac{(3,605,000 \times 144) \times \left[\dfrac{1}{12} \times \left(\dfrac{18}{12}\right)^4\right]}{10.5}}} = 0.0146 \text{ s}$$

In S.I.:

$$m = \left(\frac{457}{1,000}\right)^2 \times \frac{(23.6 \times 10^3)}{9.81} = 502.4 \text{ kg/m}$$

$$T_n = \frac{2 \times 3.1^2}{\pi\sqrt{\dfrac{(24,900 \times 10^6) \times \left[\dfrac{1}{12} \times \left(\dfrac{457}{1,000}\right)^4\right]}{502.4}}} = 0.0144 \text{ s}$$

$$t_d/T_n = 0.0054/0.0146 = 0.37$$

From ASCE/SEI Table 6.11-1, $R_{max} = 1.32$
Dynamic impact force $= R_{max}F_i = 1.32 \times 220 = 290$ kips

In S.I.:

$$t_d/T_n = 0.0055/0.0144 = 0.38$$

From ASCE/SEI Table 6.11-1, $R_{max} = 1.34$

Dynamic impact force $= R_{max}F_i = 1.34 \times 992 = 1,329$ kN

Illustrated in Fig. 3.42 is the dynamic impact force applied at $10.25/2 = 5.13$ ft (1.6 m) from the base of an exterior column, which is the critical section for flexure.

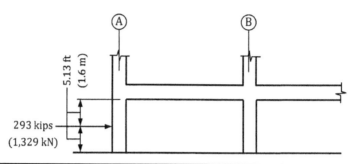

Figure 3.42 Dynamic impact force due to wood logs and poles in Example 3.2.

Exterior wall:

Assuming the wall has simple connections at both ends, the stiffness of the wall is determined as follows:

$$k = \frac{48EI}{L^3}$$

$$= \frac{48 \times (3,605,000/1,000) \times \left[\frac{1}{12} \times (32.0 \times 12) \times 10.0^3 \right]}{(11.25 \times 12)^3}$$

$$= 2,251 \text{ kip/in.} > 350 \text{ kip/in.}$$

In S.I.:

$$k = \frac{48 \times (24,900 \times 10^3) \times \left[\frac{1}{12} \times 9.8 \times \left(\frac{254}{1,000} \right)^3 \right]}{(3.4)^3}$$

$$= 406,958 \text{ kN/m} > 61,300 \text{ kN/m}$$

Therefore, use $k = 350$ kip/in. (61,300 kN/m).

Assume minimum debris weight $W_d = 1,000$ lb (4.5 kN). Debris mass, m_d:

$$m_d = W_d/g = 1,000/32.2 = 31.1 \text{ slugs}$$

In S.I.:

$$m_d = 4.5 \times 10^3/9.81 = 459 \text{ kg}$$

Nominal debris impact force, F_{ni}:

$$F_{ni} = u_{max} \sqrt{km_d} = 29.7 \times \sqrt{(350 \times 12) \times (31.1/1,000)} = 339 \text{ kips}$$

In S.I.:

$$F_{ni} = 9.1 \times \sqrt{(61,300/1,000) \times 459} = 1,526 \text{ kN}$$

For Tsunami Category II buildings, $I_{tsu} = 1.0$ from ASCE/SEI Table 6.8-1.
Orientation coefficient $C_o = 0.65$.
Design debris impact force $F_i = I_{tsu}C_oF_{ni} = 1.0 \times 0.65 \times 339 = 220$ kips
In S.I.:

$$\text{Design debris impact force } F_i = 1.0 \times 0.65 \times 1,526 = 992 \text{ kN}$$

Impulse duration for elastic impact, t_d:

$$t_d = \frac{2m_d u_{max}}{F_{ni}} = \frac{2 \times 31.1 \times 29.7}{339 \times 1,000} = 0.0054 \text{ s}$$

In S.I.:

$$t_d = \frac{2m_d u_{max}}{F_{ni}} = \frac{2 \times 459 \times 9.1}{1,526 \times 1,000} = 0.0055 \text{ s}$$

Assuming the wall has simple connections at both ends, the natural period of the wall is determined based on the natural period of an equivalent column with the width equal to one-half of the vertical span, which is $11.25/2 = 5.63$ ft (1.7 m):

$$\text{Mass of wall per unit length } m = \frac{10}{12} \times 5.63 \times \frac{150}{32.2} = 21.9 \text{ slugs/ft}$$

$$T_n = \frac{2L^2}{\pi\sqrt{\frac{EI}{m}}} = \frac{2 \times 11.25^2}{\pi\sqrt{\frac{(3,605,000 \times 144) \times \left[\frac{1}{12} \times 5.63 \times \left(\frac{10.0}{12}\right)^3\right]}{21.9}}} = 0.0318 \text{ s}$$

In S.I.:

$$m = \left(\frac{254}{1,000}\right) \times 1.7 \times \frac{(23.6 \times 10^3)}{9.81} = 1,039 \text{ kg/m}$$

$$T_n = \frac{2 \times 3.4^2}{\pi\sqrt{\frac{(24,900 \times 10^6) \times \left[\frac{1}{12} \times 1.7 \times \left(\frac{254}{1,000}\right)^3\right]}{1,039}}} = 0.0312 \text{ s}$$

$$t_d/T_n = 0.0054/0.0318 = 0.17$$

From ASCE/SEI Table 6.11-1, $R_{max} = 0.68$
Dynamic impact force $= R_{max}F_i = 0.68 \times 220 = 150$ kips

In S.I.:

$$t_d/T_n = 0.0055/0.0312 = 0.18$$

From ASCE/SEI Table 6.11-1, $R_{max} = 0.72$

Dynamic impact force $= R_{max}F_i = 0.72 \times 992 = 714$ kN

This force acts along the horizontal center of the wall.

- Vehicles

$$F_i = 30I_{tsu} = 30 \times 1.0 = 30 \text{ kips}$$

In S.I.:

$$F_i = 130I_{tsu} = 130 \times 1.0 = 130 \text{ kN}$$

This impact force is applied anywhere from 3 ft (0.91 m) up from the base of the column or wall.

- Tumbling boulders and concrete debris

This impact load must be considered because $h_{max} = 44.5$ ft (13.6 m) > 6.0 ft (1.8 m).

$$F_i = 8,000I_{tsu} = 8,000 \times 1.0/1,000 = 8 \text{ kips}$$

In S.I.:

$$F_i = 36I_{tsu} = 36 \times 1.0 = 36 \text{ kN}$$

This impact force is applied 2 ft (0.61 m) above the base of the column or wall.

- Shipping containers

Because the site is not located in an impact zone for shipping containers in accordance with ASCE/SEI 6.11.5, impact loads due to shipping containers are not applicable.

Step 8—Determine the tsunami load combinations   Sec. 3.7.2

Principal tsunami forces and effects on the lateral force–resisting system and the individual components must be combined with other specified loads using the load combinations in ASCE/SEI 6.8.3.3 [see Eqs. (3.1) and (3.2) of this publication]. As noted previously, the debris impact loads need not be combined with other tsunami-related loads determined in accordance with ASCE/SEI Chapter 6.

CHAPTER 4
References

1. International Code Council. 2017. *2018 International Building Code*, Washington, DC.
2. Structural Engineering Institute of the American Society of Civil Engineers (ASCE). 2017. *Minimum Design Loads and Associated Criteria for Buildings and Other Structures*, ASCE/SEI 7-16, Reston, VA.
3. Structural Engineering Institute of the American Society of Civil Engineers (ASCE). 2014. *Flood Resistant Design and Construction*, ASCE/SEI 24-14, Reston, VA.
4. Federal Emergency Management Agency (FEMA). FEMA Flood Map Service Center. https://msc.fema.gov/portal/home.
5. American Society of Civil Engineers (ASCE). ASCE 7 Hazard Tool. https://asce7hazardtool.online/.
6. Federal Emergency Management Agency (FEMA). 2014. *Guidance for Flood Risk Analysis and Mapping—Vertical Datum Conversions*, Washington, DC.
7. Federal Emergency Management Agency (FEMA). 2011. *Coastal Construction Manual: Principles and Practices of Planning, Siting, Designing, Constructing, and Maintaining Residential Buildings in Coastal Areas, 4th Edition*, FEMA P-55, Washington, DC.
8. Federal Emergency Management Agency (FEMA). 2012. *Engineering Principles and Practices of Retrofitting Floodprone Residential Structures, 3rd Edition*, FEMA P-259, Washington, DC.
9. Federal Emergency Management Agency (FEMA). Wave Height Analysis for Flood Insurance Studies (WHAFIS), Version 4.0, https://www.fema.gov/wave-height-analysis-flood-insurance-studies-version-40, Washington, DC.
10. US Army Corps of Engineers (USACE). 1995. *Flood Proofing Regulations*, Washington, DC.
11. PIANC. 2014. *Harbor Approach Channels: Design Guidelines*, Appendix C, Typical Ship Dimensions, Report No. 121-2014, Brussels, Belgium.